Stability and Reactivity of Crown-Ether Complexes

Stability and Reactivity of Crown-Ether Complexes

F. DE JONG
Koninklijke/Shell Laboratorium
Amsterdam, The Netherlands

D. N. REINHOUDT
Department of Chemical Technology
Twente University of Technology
Enschede, The Netherlands

Reprinted from Advances in Physical Organic Chemistry Volume 17

1981

ACADEMIC PRESS
A Subsidiary of Harcourt Brace Jovanovich, Publishers
London • New York • Toronto • Sydney • San Francisco

ACADEMIC PRESS INC. (LONDON) LTD.
24/28 Oval Road
London NW1

United States edition published by
ACADEMIC PRESS INC.
111 Fifth Avenue
New York, New York 10003

British Library Cataloguing in Publication Data

Jong, F. de
Stability and reactivity of crown-ether complexes.
1. Ethers
I. Title II. Reinhoudt, D. N.
547′.035 QD305.E7

ISBN 0–12–208780–1

LCCN 81–67897

Printed in Great Britain by Spottiswoode Ballantyne Ltd., Colchester and London

Preface

The selectivity and high efficiency of biological processes are increasingly the source of inspiration for research. Mimicry of such processes in the laboratory and eventually in industry is an important objective in current chemistry. Mutual recognition of species at a molecular level is essential in such processes.

Their ability to form well-defined complexes with salts and neutral molecules may explain the exponential growth of the interest in synthetic macrocyclic ligands. Crown-ethers, cryptands and acyclic polyethers are now being used in synthetic and analytical chemistry, in mechanistic studies, and in applied chemistry.

In the first period following the discovery of the crown-ethers by Pedersen in 1967, the emphasis of the efforts was on the synthesis and on qualitative complexation studies. This work has been extensively reviewed as mentioned in chapter 1.

The physical chemistry of the complexation phenomena and a quantitative understanding of the effects of complexation on the reactivity of complexed species is very important for a proper application of macrocyclic ligands.

The aim of this book is a critical evaluation of these aspects of crown-ether chemistry; only cyclic compounds having at least one oxygen donor atom have been included. Literature up to the end of 1978 is covered.

In the first chapter the scope and limitations of this book are outlined and leading references to reviews on other aspects of crown-ether chemistry are given. The second chapter deals with complexation of metal cation salts and the parameters determining the thermodynamic stability. Chapter 3 describes the kinetics of metal cation complexation and in chapter 4 the chemical reactivity of such complexes are discussed. Nucleophilic substitution and addition, hydrogen abstraction by crown-ether solubilized bases including carbene generation, oxidation and reduction reactions are the major reaction types that have been reviewed.

In chapter 5 the properties of complexes with ammonium salts, particularly their kinetic and thermodynamic stabilities are summarized. Complexes of chiral crown ethers with racemic ammonium salts are described separately in chapter 6 which includes a section on enzyme-type of catalysis.

Chapter 7 describes complexation of arenediazonium salts with the emphasis on the influence of complexation on the chemical reactivity. The book concludes with a chapter on complexes of crown-ethers with neutral organic molecules.

We hope that this monograph will be of assistance to all who are actively involved in crown-ether chemistry but also to those who are using crown-ethers incidentally as tools in their research.

F. de Jong
D. N. Reinhoudt

Amsterdam/Enschede 1981

Contents

1 Introduction

In 1967 Pedersen reported on his pioneering work on the synthesis of macrocyclic polyethers and his observation that these "crown ethers", as he called them, were capable of forming surprisingly stable complexes with alkali- and alkaline-earth metal cations (Pedersen, 1967a,b). His finding aroused a great deal of interest in many areas of chemistry and this interest is still growing. The preparation of these "simple", first-generation crown ethers—having only oxygen donor atoms—was rapidly followed by the synthesis of ligands having nitrogen and sulfur as (additional) donor atoms. Monocyclic ligands, in turn, were succeeded by bicyclic compounds, such as Lehn's cryptands (Lehn, 1973, 1978). The introduction of additional binding sites or steric barriers at strategic positions near the cavity of the crown ether resulted in shape-selective ligands (Cram and Cram, 1974, 1978), and the mimicry of enzyme systems (Chao and Cram, 1976). Extensive reviews on the synthesis of such compounds have been published recently (Bradshaw and Hui, 1974; Bradshaw, 1978; Bradshaw *et al.*, 1979a; Christensen *et al.*, 1974; Newkome *et al.*, 1977; Pedersen, 1978; Pedersen and Frensdorf, 1972; Reinhoudt and De Jong, 1979; Vögtle and Neumann, 1973; Harris, 1977).

The unique ability of crown ethers to form stable complexes with various cations has been used to advantage in such diverse processes as isotope separations (Jepson and De Witt, 1976), the transport of ions through artificial and natural membranes (Tosteson, 1968) and the construction of ion-selective electrodes (Ryba and Petranek, 1973). On account of their lipophilic exterior, crown ether complexes are often soluble even in apolar solvents. This property has been successfully exploited in liquid–liquid and solid–liquid phase-transfer reactions. Extensive reviews deal with the synthetic aspects of the use of crown ethers as phase-transfer catalysts (Gokel and Dupont Durst, 1976; Liotta, 1978; Weber and Gokel, 1977; Starks and Liotta, 1978). Several studies have been devoted to the identification of the factors affecting the formation and stability of crown-ether complexes, and many aspects of this subject have been discussed in reviews (Christensen *et al.*, 1971, 1974; Pedersen and Frensdorf, 1972; Izatt *et al.*, 1973; Kappenstein, 1974).

The present article will be concerned mainly with the physical organic aspects of crown-ether chemistry. Since the effect of crown ethers on the characteristics of organic reactions depends greatly on complexation phenomena, we will first discuss the thermodynamic and kinetic stabilities of crown-ether complexes before dealing with the chemical reactivity of the complexes themselves. In this review the term macrocyclic polyethers is used mainly to refer to cyclic compounds having at least one oxygen donor atom. Open-chain polyethers, octopus molecules (Vögtle and Weber, 1974), polypode ligands (Fornasier *et al.*, 1976), all-sulfur- (Bradshaw and Hui, 1974) and all-nitrogen-containing (Busch *et al.*, 1971; Busch, 1978) macroheterocycles are included only if sufficient data are available that allow a comparison with crown ethers. The same holds for cryptands for which excellent review articles are available (Lehn, 1973, 1978). The effect of crown ethers on the ion-pairing properties of carbanions has been ignored deliberately in this review because it has been discussed in a previous volume of this series (Hogen-Esch, 1977). Finally, to limit the scope of this review even further, we have disregarded radical anions (except superoxide) and solvated electrons, crown ether-assisted photochemical reactions, metal-separation processes, the transport of cations through liquid and polymeric membranes, anionic polymerizations, and spectroscopic properties of crown ether complexes. Literature up to the end of 1978 is covered.

2 Complexation of metal cations

GENERAL

In a solution containing a crown ether (Cr) and a salt of a univalent cation (MX), the equilibria (1) can exist. D_{ip} and D_{csip} are the dissociation constants

$$\begin{array}{ccc} M^+.X^- & \overset{D_{ip}}{\rightleftarrows} & M^+ + X^- \\ + & & + \\ Cr & & Cr \\ K_{ip}\updownarrow & & \updownarrow K_i \\ M^+.Cr.X^- & \underset{D_{csip}}{\rightleftarrows} & M^+.Cr + X^- \end{array} \qquad (1)$$

of the ion pair $M^+.X^-$ and the crown ether-separated ion pair $M^+.Cr.X^-$ respectively. K_i and K_{ip} are the association constants of the free cation M^+ and the ion pair $M^+.X^-$ respectively with the crown ether. Since $K_{ip} \times D_{csip} = K_i \times D_{ip}$, the system can be described by three parameters. In polar solvents such as water and methanol, D_{ip} is very high for most salts, and consequently K_i is the only association constant of importance. In apolar solvents, on the other hand, one deals mainly with K_{ip}.

The association constants K_i and K_{ip} have been determined by various methods. Izatt and co-workers (Izatt *et al.*, 1971, 1976a, 1976b, 1977a, 1977b, 1977c, 1978a) have determined K_i in water and methanol by calorimetric titration. The introduction of a cation into the cavity of the crown ether frequently leads to marked changes in the properties of the crown moiety. Thus UV spectroscopy (Hautala and Hastings, 1978), fluorescence spectroscopy

(Platzner *et al.*, 1978), and ^{1}H- and ^{13}C-nmr spectroscopy (Live and Chan, 1976) have been used to determine association constants. In the case of an optically active crown ether, circular dichroism can be used (Mack *et al.*, 1976). The different properties of the cation in M^+ and $M^+.Cr$ (or $M^+.X^-$ and $M^+.Cr.X^-$) allow K_i (or K_{ip}) determinations by conductometric (Evans *et al.*, 1972), potentiometric (Frensdorf, 1971a), and electrochemical (Gordon and Rock, 1977) methods, as well as by nmr spectroscopy of the cation: ^{23}Na (Shchori and Jagur-Grodzinsky, 1973), ^{39}K (Shporer and Luz, 1975), ^{133}Cs (Mei *et al.*, 1977a) and ^{205}Tl (Srivanavit *et al.*, 1977). For some organic anions, such as picrate and fluorenyl, the UV absorption of the anion in the tight ion pair $M^+.X^-$ is quite different from that in the crown ether-separated ion pair $M^+.Cr.X^-$. Smid (1972, 1976) has made extensive use of this property.

The complexing ability of crown ethers in solvents of low polarity has been studied using two-phase partition experiments (Frensdorf, 1971b). The equilibrium between an aqueous solution of the salt (MX) and an organic solution containing the crown ether (Cr) is given by (2). Further dissociation of

$$M^+_{aq} + X^-_{aq} + Cr_{org} \underset{}{\overset{K_{ext}}{\rightleftharpoons}} M^+.Cr.X^-_{org} \tag{2}$$

the crown ether-separated ion pair may occur according to (1). The extraction constant K_{ext} and the association constant K_{ip} are related (Smid, 1976; Moore *et al.*, 1977) through (3) in which K_D, the distribution constant in the absence

$$K_{ext} = K_D . K_{ip} \tag{3}$$

of ligand (4), can be determined separately, for instance by UV spectroscopy (Moore *et al.*, 1977) or radioactive labelling (Haynes and Pressman, 1974). In

$$M^+_{aq} + X^-_{aq} \overset{K_D}{\rightleftharpoons} M^+.X^-_{org} \tag{4}$$

the case of water-soluble ligands, the partition of the crown ether over the two phases and the association equilibria in the aqueous phase should also be considered. This leads to a more complicated scheme (Marcus and Asher, 1978).

Some caution is required when comparing the association constants obtained from extraction experiments with those measured under anhydrous, homogeneous conditions. Iwachido *et al.* (1976, 1977) have shown that the extracted cation retains part of its aqueous solvation shell on complexation. In particular, the small univalent cations (Li^+, Na^+) and bivalent cations give high hydration numbers for their crown-ether complexes. Water molecules completing the co-ordination sphere of the cation have frequently been encountered in the solid state of crown-ether complexes (Bush and Truter, 1970, 1971). The effect of small amounts of water on the equilibria (1) has not been studied yet for crown ethers. However, it has been found that the presence of small amounts of water in acetone causes the complex-formation constants of Na^+, K^+, and Cs^+ with the macrocyclic antibiotic nonactin to drop from 7×10^4, 7×10^4, 1×10^4 to 210, 2×10^4, and 400, respectively (Prestegard and Chan, 1970).

Another type of biphase partition comprises equilibration of a crown ether-containing solution with a sparingly soluble solid salt. Assuming that the amount of free salt [$M^+.X^-$ in (1)] at equilibrium equals the solubility—which can be determined separately—one can calculate the association constant K_{ip} from the amount of solubilized salt. Reinhoudt *et al.* (1977) applied this technique using Zeise's salts.

The stability of crown-ether complexes depends on several factors; these include cavity size of the ligand, cation diameter, spatial distribution of ring binding sites, the character of the hetero-atoms, the presence of additional binding sites and the type of solvent used. In apolar solutions it also depends on the nature of the anion. The effects of these parameters will be illustrated in the next sections.

CAVITY SIZE AND CATION DIAMETER

Pedersen's (1967a) original model, in which the cation is captured in the cavity of the crown ether, has been fully substantiated by X-ray crystallographic studies (Truter, 1973; Dalley, 1978). On the basis of this model, the main features of cation binding can be explained surprisingly well. The size of the cavity can be estimated from CPK molecular models (Table 1). The binding of

TABLE 1

Estimated cavity sizes (nm)

Ligand	CPK molecular models[a]	X-ray[b]
12-crown-4	0.12	
15-crown-5	0.17	0.17–0.18
18-crown-6	0.26	0.25–0.28
21-crown-7	0.34	

[a] Pedersen, 1967a
[b] Dalley, 1978

several cations in water by (*cis-syn-cis*)-dicyclohexyl-18-crown-6 [20], having an estimated cavity size of about 0.26 nm, is shown in Fig. 1. Within the same category of (univalent or bivalent) cations, the most stable complexes are formed when the ratio of cation and cavity diameters is close to unity. Any deviation from the perfect fit results in a loss of binding capacity.

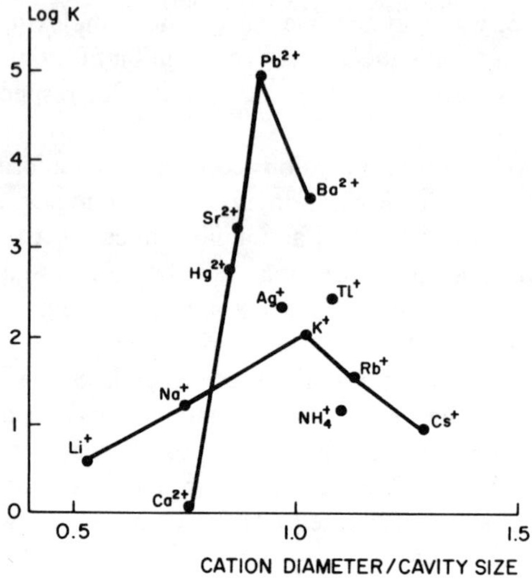

FIG. 1 Relation between log K and the ratio of cation diameter to cavity size for (*cis-syn-cis*)-dicyclohexyl-18-crown-6 [20] in water at 25°C. Ionic diameters taken from Pedersen (1967a), and log K values from Izatt *et al.* (1976a) and Frensforf (1971a)

The remarkable stability of crown ether complexes with uni- and bivalent cations is enthalpic in origin (Izatt *et al.*, 1976a). The enthalpy and entropy contributions to the free energy of binding for complexes of 18-crown-6 [3] in water are given in Table 2. Among the alkali cations, there is a preference for K^+ owing to its favourable enthalpy of binding. The K^+/Tl^+ and K^+/Ag^+ selectivities depend on both enthalpic and entropic factors. Considering the close similarity between the diameters of the K^+ (0.266 nm) and Ag^+ ions

TABLE 2

Enthalpy and entropy contributions (in kcal mol^{-1})[a] to the binding of cations by 18-crown-6 [3] in water at 25°C[b]

Cation	$-\Delta H$	$T\Delta S$	Cation	$-\Delta H$	$T\Delta S$
Na^+	2.25	−1.10			
K^+	6.21	−3.40	Hg^{2+}	4.69	−1.40
Rb^+	3.82	−1.70	Sr^{2+}	3.61	+0.09
Cs^+	3.79	−2.41	Pb^{2+}	5.16	+0.66
Tl^+	4.44	−1.34	Ba^{2+}	7.58	−2.35
Ag^+	2.17	−0.12			
NH_4^+	2.34	−0.66			

[a] 1 kcal mol^{-1} = 4.2 kJ mol^{-1}
[b] Izatt *et al.*, 1976a

(0.252 nm), the relatively low $-\Delta H$ value for the latter probably reflects its "soft" character. For bivalent cations complex formation is favourable from an enthalpy point of view, with either favourable (Sr^{2+}, Pb^{2+}) or unfavourable (Hg^{2+}, Ba^{2+}) entropy contributions.

The association constant K_1 for several 1:1 complexes of crown ethers [1]–[23] have been determined in water (Table 3). Compounds having an

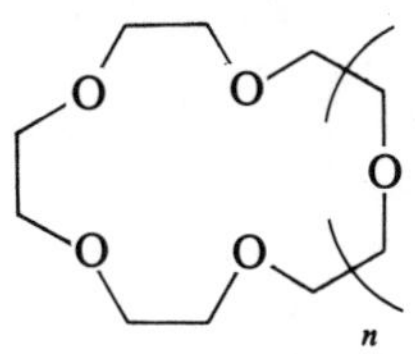

[1], $n = 0$
[2], $n = 1$
[3], $n = 2$
[4], $n = 3$
[5], $n = 4$

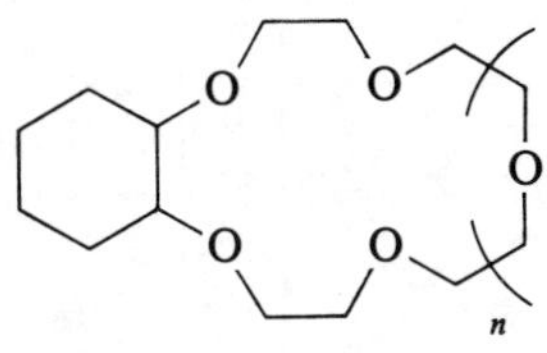

[6], $n = 1$
[7], $n = 2$

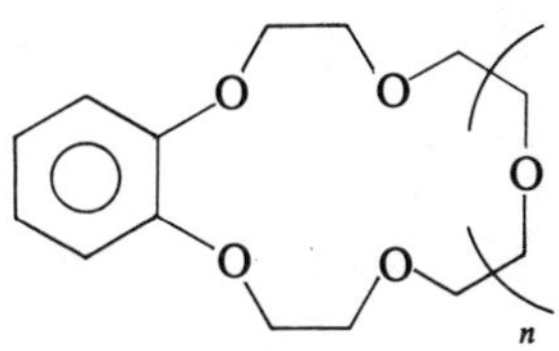

[8], $n = 1$
[9], $n = 2$
[10], $n = 3$

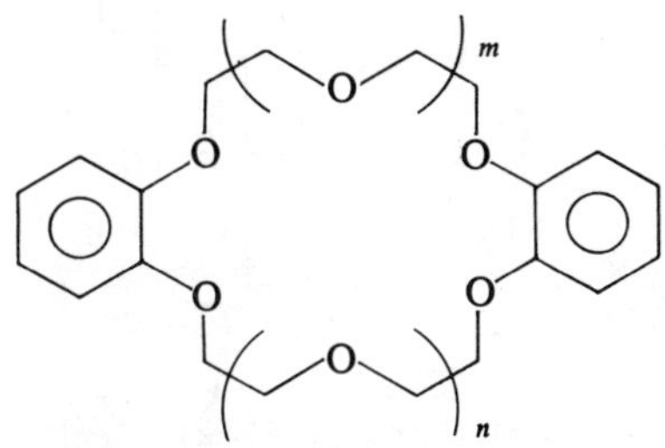

[11], $m = n = 1$
[12], $m = 1, n = 2$
[13], $m = n = 2$
[14], $m = n = 3$
[15], $m = n = 8$

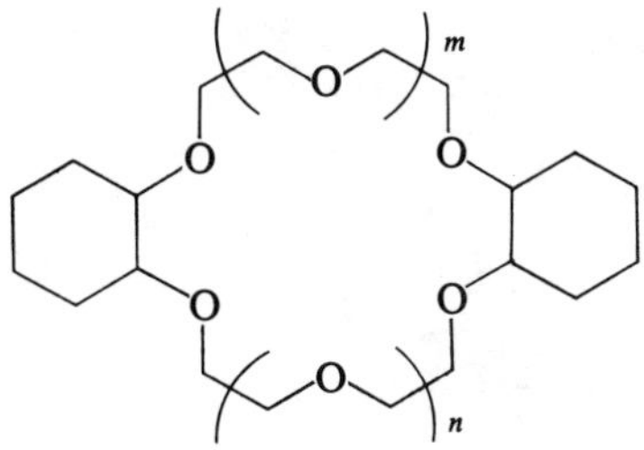

[16], $m = n = 0$
[17], $m = 0, n = 1$
[18], $m = 2, n = 1$
[19], $m = n = 2$

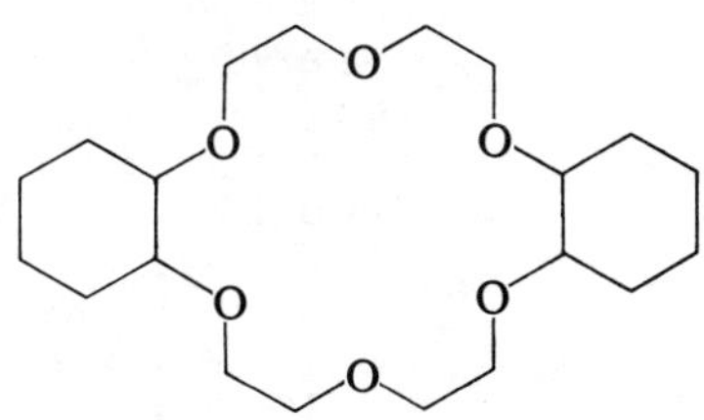

[20], *cis-syn-cis*
[21], *cis-anti-cis*
[22], *trans-syn-trans*
[23], *trans-anti-trans*

TABLE 3

Log K_1 values for 1 : 1 complexes in water at 25 °C

Ligand	Li^+	Na^+	K^+	Rb^+	Cs^+	Ag^+	Tl^+	NH_4^+	Sr^{2+}	Ba^{2+}	Pb^{2+}	Hg^{2+}
15-Crown-5 [2]	—	0.70[a]	0.74[a]	0.62[a]	0.8[a]	0.94[a]	1.23[a]	1.71[a]	1.95[a]	1.71[a]	1.85[e]	1.68[a]
Cyclohexyl-15-crown-5 [6]	<1.0[b]	—	0.6[b]	—	—	—	—	—	—	—	—	—
Benzo-15-crown-5 [8]	—	0.4[d]	0.38[d]	—	—	—	—	—	—	—	—	—
18-Crown-6 [3]	—	0.80[a]	2.03[a]	1.56[a]	0.99[a]	1.50[a]	2.27[a]	1.23[a]	2.72[a]	3.78[a]	4.27[e]	2.42[a]
Cyclohexyl-18-crown-6 [7]	<0.7[b]	0.8[b]	1.9[b]	—	0.8[b]	1.8[b]	—	1.1[b]	—	—	—	—
Dicyclohexyl-18-crown-6												
cis-syn-cis [20]	0.6[b]	1.21[a]	2.02[c]	1.52[c]	0.96[c]	2.36[a]	2.44[a]	1.33[c]	3.24[c]	3.57[c]	4.95[a]	2.75[a]
cis-anti-cis [21]	—	0.69[a]	1.63[c]	0.87[c]	0.9[b]	1.59[c]	1.83[a]	0.80[c]	2.64[c]	3.27[c]	4.43[a]	2.60[a]
Benzo-18-crown-6 [9]	—	—	2.0[f]	—	—	—	—	—	—	3.7[f]	—	—
Dibenzo-18-crown-6 [11]	—	1.1[g]	1.6[g]	1.1[g]	0.83[h]	1.4[h]	1.5[h]	~0.3[h]	1.0[h]	1.95[h]	1.89[h]	—
Dicyclohexyl-21-crown-7 [18]	—	—	—	—	1.9[b]	—	—	—	—	—	—	—
Dicyclohexyl-24-crown-8 [19]	—	—	—	—	1.9[b]	—	—	—	—	—	—	—

[a] Izatt *et al.*, 1976a
[b] Frensdorf, 1971a
[c] Izatt *et al.*, 1971
[d] Izatt *et al.*, 1976b
[e] Izatt *et al.*, 1978a
[f] Live and Chan, 1976
[g] Harris, 1977
[h] Shchori *et al.*, 1975

18-crown-6-type structure (such as [3], [7], [9], [11], and [21]) show a cation selectivity similar to that of [20], as shown in Fig. 1. The binding constants of 15-crown-5 are much lower than those of 18-crown-6 for all cations except Na^+ and NH_4^+. The larger polyethers dicyclohexyl-21-crown-7 [18] and dicyclohexyl-24-crown-8 [19] show an increased ability to bind Cs^+.

For crown ethers containing 24-membered rings and larger, the binding capacities are difficult to estimate on the basis of the "ion-in-the-hole" model. Owing to their high flexibility, the cavity sizes cannot be defined properly. Table 4 gives the log K_1 values for several 1 : 1 complexes in methanol, and it is seen that crown ethers as large as dibenzo-30-crown-10 [14] and dibenzo-60-crown-20 [15] form remarkably stable complexes with K^+ (Frensdorf, 1971a) and Tl^+ (Hofmanova *et al.*, 1978). An X-ray study of the [14].KI

TABLE 4

Log K_1 values for 1 : 1 complexes in methanol at 25°C

Ligand	Na^+	K^+	Rb^+	Cs^+	Ag^+	Tl^+	NH_4^+	Ba^{2+}
Dicyclohexyl-12-crown-4 [16]	2.18[a]	1.30[a]	—	—	—	—	—	—
Cyclohexyl-15-crown-5 [6]	3.71[a]	3.58[a]	—	2.78[a]	—	—	—	—
Dicyclohexyl-15-crown-5 [17]	—	—	—	—	—	—	—	—
Benzo-15-crown-5 [8]	—	—	—	—	—	3.00[c]	—	—
18-Crown-6 [3]	4.32[a]	6.10[a]	5.35[b]	4.62[a]	4.58[g]	—	4.27[f]	7.0[d]
Cyclohexyl-18-crown-6 [7]	4.09[a]	5.89[a]	—	4.30[a]	—	—	—	—
Dicyclohexyl-18-crown-6								
cis-syn-cis [20]	4.08[a]	6.01[a]	—	4.61[a]	—	} 5.20[c]	—	—
cis-anti-cis [21]	3.68[a]	5.38[a]	—	3.49[a]	—		—	—
trans-syn-trans [22]	2.99[b]	4.14[b]	3.42[b]	3.00[b]	—	—	—	—
trans-anti-trans [23]	2.52[b]	3.26[b]	2.73[b]	2.27[b]	—	—	—	—
Benzo-18-crown-6 [9]	—	5.20[c]	—	—	—	4.60[c]	—	—
Dibenzo-18-crown-6 [11]	4.36[a]	5.00[a]	—	3.55[a]	—	4.00[c]	—	—
	—	4.60[c]	—	—	—	3.62[e]	—	—
21-Crown-7 [4]	—	4.41[a]	—	5.02[a]	—	—	—	—
Dibenzo-21-crown-7 [12]	2.40[a]	4.30[a]	—	4.20[a]	—	—	—	—
24-Crown-8 [5]	—	3.48[a]	—	4.15[a]	—	—	—	—
Dibenzo-24-crown-8 [13]	—	3.49[a]	—	3.78[a]	—	3.40[c]	—	—
	—	3.20[c]	—	—	—	—	—	—
Dibenzo-30-crown-10 [14]	2.0[a]	4.60[a]	—	—	—	4.10[c]	—	—
	—	3.90[c]	—	—	—	—	—	—
Dibenzo-60-crown-20 [15]	—	3.90[a]	—	—	—	—	—	—

[a] Frensdorf, 1971a
[b] Burden *et al.*, 1977; Coxon *et al.*, 1978
[c] Hofmanova *et al.*, 1978
[d] Izatt *et al.*, 1977a
[e] Platzner *et al.*, 1978

complex revealed that the crown ether is wrapped around the cation to form a three-dimensional cavity with all ten oxygen atoms co-ordinated to the cation (Bush and Truter, 1970, 1971). From NMR studies of complexes of large macrocyclic polyethers in chloroform solution, Reinhoudt *et al.* (1977) concluded that the macrocycle adjusts the size of the cavity to the size of the cation. This twisting leads to a two-dimensional cavity of reduced size.

Crown ethers whose cavity is too small to accommodate the anion overcome this handicap in an ingenious way by forming sandwich-type complexes (5). The formation of 2:1 complexes is very sensitive to steric

$$M^+ + Cr \underset{}{\overset{K_1\,(1:1)}{\rightleftarrows}} M^+.Cr + Cr \underset{}{\overset{K_1\,(2:1)}{\rightleftarrows}} M^+.(Cr)_2 \qquad (5)$$

$$K_1\,(1:1) = \frac{[M^+.Cr]}{[M^+][Cr]} \qquad K_1\,(2:1) = \frac{[M^+.(Cr)_2]}{[Cr][M^+.Cr]}$$

repulsions of the two crown moieties in the complex. As follows from Table 5, benzo-crown ethers form more stable 2:1 complexes than the corresponding cyclohexyl or parent compounds; benzo-15-crown-5 [8] forms particularly stable 2:1 complexes with K^+ and Rb^+ (Izatt *et al.*, 1976b). In the solid state, the benzo-15-crown-5 complex with K^+ is also of the sandwich type (Mallinson and Truter, 1972). Clear evidence of the proximity of the ligands was recently presented by Ishizu *et al.* (1978), who studied the effect of complexation on the esr spectrum of the spin-labelled benzo-15-crown-5 [24]

TABLE 5

Log K_1 values for 2:1 complexes at 25°C

Ligand	Cation	Solvent[a]	Log K_1 (2:1)	Log $\frac{K_1\,(2:1)}{K_1\,(1:1)}$
Cyclohexyl-15-crown-5 [6]	K^+	M	1.88[b]	−1.7
	Cs^+	M	1.91[b]	−0.89
Benzo-15-crown-5 [8]	K^+	M/W	2.65[c]	+1.15
	Rb^+	M/W	1.97[c]	+0.17
Dibenzo-18-crown-6 [11]	Cs^+	M	2.92[b]	−0.64
18-Crown-6 [3]	Cs^+	M	1.30[b]	−3.3
Cyclohexyl-18-crown-6 [7]	Cs^+	M	1.52[b]	−2.7
Dicyclohexyl-18-crown-6 *cis-syn-cis* [20]	Cs^+	M	0.59[b]	−4.0
Dibenzo-21-crown-7 [12]	Cs	M	1.9[b]	−2.3

[a] M, methanol; M/W, methanol/water (70/30)
[b] Frensdorf, 1971a
[c] Izatt *et al.*, 1976b

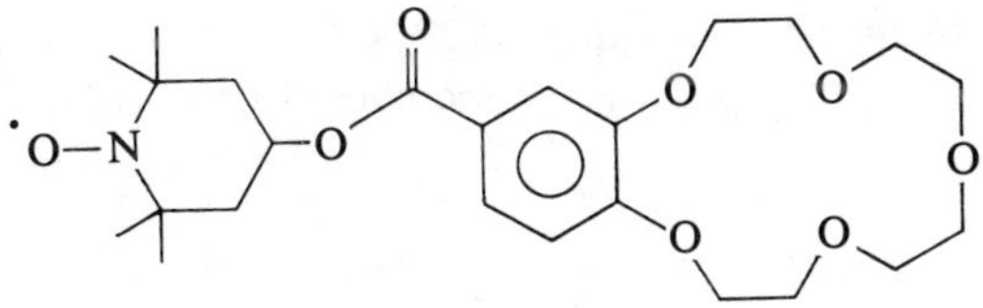

[24]

in ethanol solution. In the presence of Na^+ the esr spectrum hardly differed from that of the uncomplexed parent compound, whereas in the presence of K^+ a strong spin–spin interaction was observed.

The formation of complexes of 2:1 stoichiometry is not restricted to cyclic ligands; it has also been observed with linear polyethers (Kodama and Kimura, 1976; Tümmler *et al.*, 1977).

SPATIAL ARRANGEMENT OF BINDING SITES

From an electrostatic point of view, the crown-ether complexes derive their binding energy mainly from ion–dipole interactions. In the most ideal two-dimensional situation, all dipoles of the CH_2OCH_2 moiety are directed towards the centre of the cavity. Any changes in number, direction and magnitude of these dipoles are expected to show up in the binding constants. A detailed discussion of these factors for four of the five possible dicyclohexyl-18 crown-6 isomers has been given by Burden *et al.* (1977) and Coxon *et al.* (1978). Their data and those reported by Frensdorf (1971a) show the following order of complexing ability towards alkali cations (Table 4): *cis-syn-cis* > *cis-anti-cis* > *trans-syn-trans* > *trans-anti-trans*. The enthalpy of complex formation was found to be more favourable for the *cis-anti-cis* than for the *cis-syn-cis* isomer, but a less unfavourable entropy contribution for the latter resulted in the above order of free energies (Izatt *et al.*, 1971).

Drastic changes in complexing ability are observed upon replacing a CH_2CH_2 unit in 18-crown-6 by other groups. The binding constants for K^+ in methanol were 1×10^4 and 2.7×10^4 times lower for compounds [25] and [26]

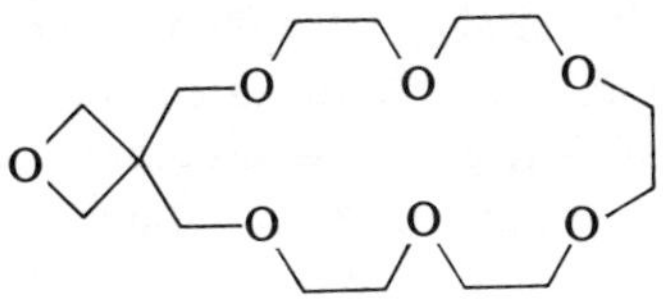

[25]

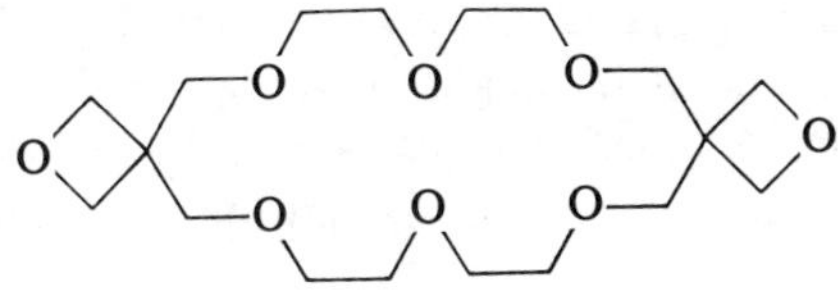

[26]

than for 18-crown-6 (Krespan, 1974). The effect of aromatic and hetero-aromatic units on the complexing ability towards Zeise's salt has been studied in chloroform (Reinhoudt *et al.*, 1977; Kawashima *et al.*, 1978) for compounds [27]–[35]. Among the 17-membered rings ([27], [33], [34], [35],

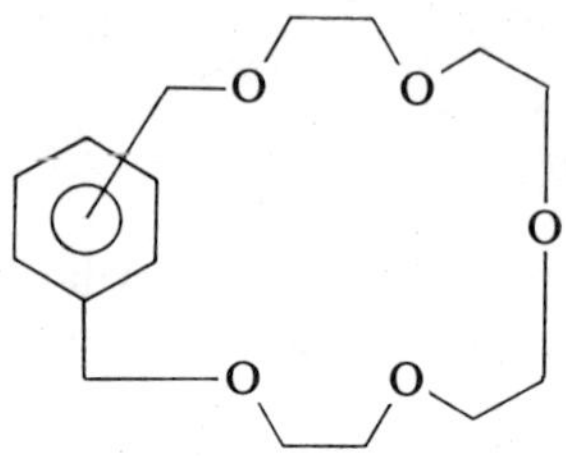

[27], *ortho*

[28], *meta*

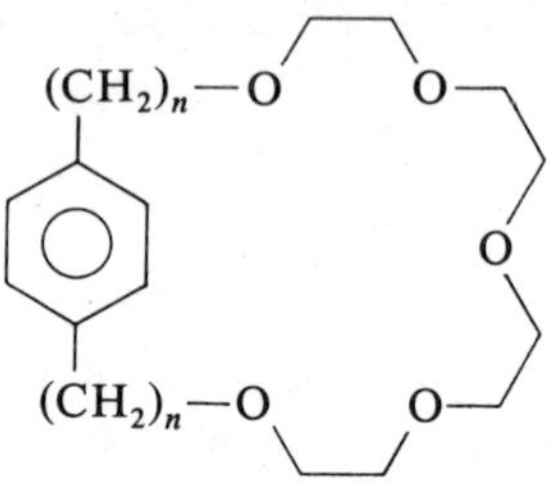

[29], $n = 0$

[30], $n = 1$

[31], $n = 2$

[32], $n = 3$

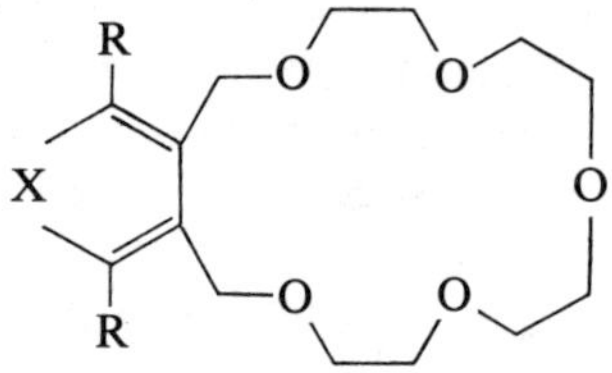

[33], X = O, R = H

[34], X = S, R = CH_3

[35], X = SO_2, R = CH_3

and [29]) the thieno- and furo-compounds form more stable complexes than the corresponding benzo-compound. Oxidation of the sulfur atom decreases the stability of the complex. Among the paracyclophanes, the 21-membered compound shows the strongest affinity for K^+. It has been suggested (Kawashima *et al.*, 1978; Frensch and Vögtle, 1977) that suitably placed aromatic units might serve as a π-donor binding site. Although the data in Table 6 indeed suggest a dependence on the character of the aromatic unit, further studies must be awaited in which conformational effects have been minimized.

The study of Lehn's cryptands has shown that a three-dimensional arrangement of binding sites leads to very stable inclusion complexes (cryptates) with many cations. For example, the stability constant for K^+ in methanol/water (95/5) is five orders of magnitude higher with [2.2.2]-cryptand [37] (log K_i: 9.75; Lehn and Sauvage, 1975) than with [2.2]-cryptand [38] (log

TABLE 6

Log K_{ip} values for $K[PtCl_3(C_2H_4)]$ complexes of crown ethers of [27]–[35] in $CDCl_3$ at 27°C

Ligand	Ring size	Log K_{ip}	Ligand	Ring size	Log K_{ip}
[27]	17	5.85[a]			
[28]	18	6.23[a]	[29]	17	4.99[b]
[30]	19	5.41[b]	[30]	19	5.41[b]
			[31]	21	>6.7[b]
[33]	17	>6.7[a]	[32]	23	4.87[b]
[34]	17	>6.7[a]			
[35]	17	5.52[a]			

[a] Reinhoudt *et al.*, 1977
[b] Calculated from the data of Kawashima *et al.* (1978), using 4×10^{-6} M for the solubility of Zeise's salt

K_i: 4.38; Dietrich *et al.*, 1973; Lehn and Simon, 1977). This macrobicyclic effect has been shown to be of enthalpic origin (Kaufmann *et al.*, 1976; Anderegg, 1975).

[36], $n = 0$

[37], $n = 1$

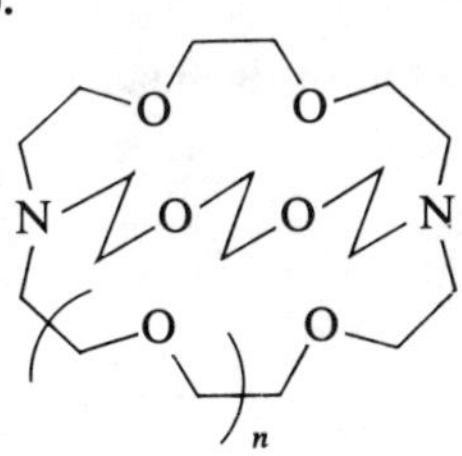

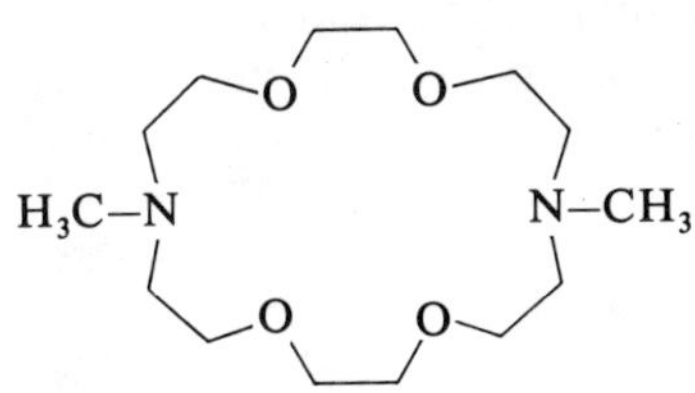

[38]

Attracted by the benefits of the cryptand effect, Coxon and Stoddart (1975, 1977) prepared the bicyclic compound [39]. Compared with the monocyclic crown ether [40], the bicyclic compound showed no increased ion binding

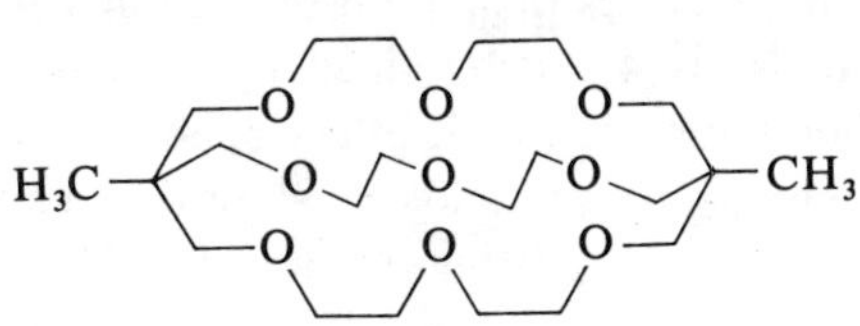

[39]

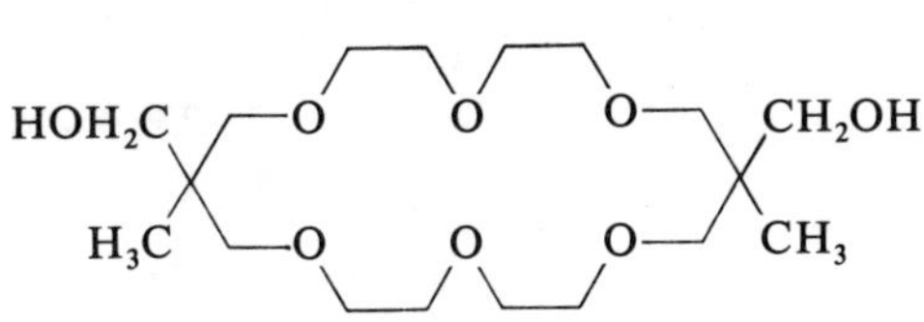

[40]

TABLE 7

Macrocyclic (MC)[a] and macrobicyclic (MBC)[b] effects

	Ligand	Solvent[c]	Na^+	K^+	Cs^+
MC	Cyclohexyl-15-crown-5 [6]	M	2.43	1.86	1.33
MC	18-Crown-6 [3]	M	2.85	3.90	2.77
MC	21-Crown-7 [4]	M	—	1.86	2.85
MC	24-Crown-8 [5]	M	—	0.61	1.74
MBC	[39]	M	−0.1	0.2	−0.2
MBC	[41]	W	4.3	4.2	—
MBC	[2.2.2]-Cryptand [37]	M/W (95/5)	3.9	5.4	−1.6

[a] log K (monocyclic) − log K (glyme)
[b] log K (bicyclic) − log K (monocyclic)
[c] M, methanol; W, water

capacity (Table 7). It was concluded that the introduction of a bridge did not compensate for the unfavourable effect of the $OCH_2CR_2CH_2O$ units. The macrobicyclic compounds [41] and [42], prepared by Parsons (1978), have only one such unfavourable fragment at the bridgehead position and show a large macrobicyclic effect (Table 7).

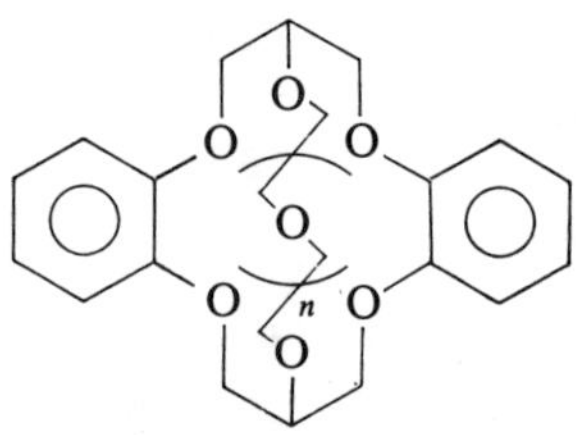

[41], $n = 2$

[42], $n = 3$

For monocyclic crown ethers the data presented in Table 4 and the stability constants for glymes [43]–[46] determined by Chaput *et al.* (1975) can be combined to calculate the macrocyclic effect (Table 7). The data indicate that the gain in binding energy on ring closure shows the same pattern as the ion selectivity of the crown ether, being highest for Na^+/15-crown-5, K^+/18-

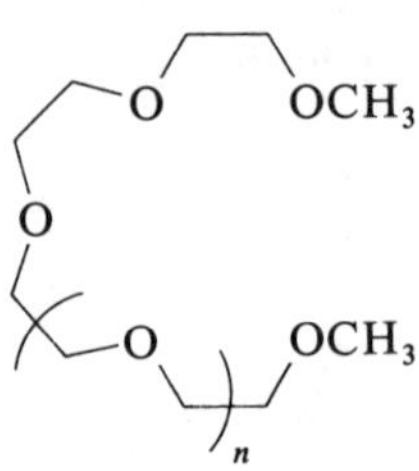

[43], $n = 1$

[44], $n = 2$

[45], $n = 3$

[46], $n = 4$

crown,-6, and Cs^+/21-crown-7 combinations. The few data available indicate that the macrocyclic effect contains a large entropy factor (Kodama and Kimura, 1976; Grandjean *et al.*, 1978).

ADDITIONAL BINDING SITES

The aim of introducing additional binding sites at strategic positions near the cavity of the crown ether is to create a three-dimensional array of binding sites without calling in the help of the macrobicyclic effect. Gramain and Frère (1979) have studied the effect of two additional OH sites on the complexing

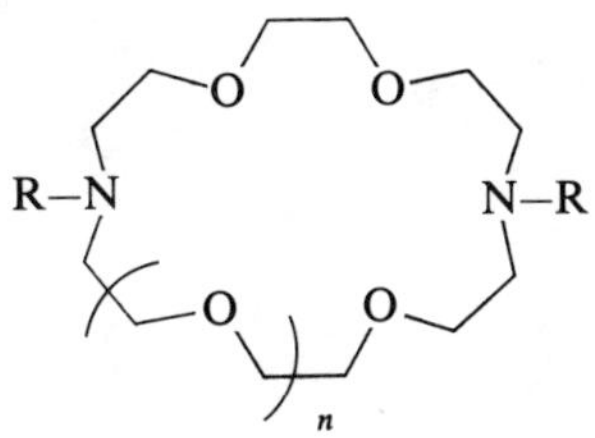

[47], $n = 0$, $R = CH_3$

[48], $n = 0$, $R = CH_2CH_2OH$

[38], $n = 1$, $R = CH_3$

[49], $n = 1$, $R = CH_2CH_2OH$

ability of diaza-crown ethers [38] and [47]–[49] in water. Their results (Table 8) show that the stability constants for complexes of diaza-15-crown-5 [47] with bivalent cations such as Ca^{2+}, Sr^{2+} and Ba^{2+} increase by factors of 85, 89 and 30 respectively upon introduction of two hydroxyl groups in the side chain. Qualitatively, the same effect was observed for the diaza-18-crown-6

TABLE 8

Effect of additional binding sites on log K_1 values in water at 20°C

Ligand	Na^+	K^+	Mg^{2+}	Ca^{2+}	Sr^{2+}	Ba^{2+}	Ref.
[47]; $n = 0$, $R = CH_3$	$\leqslant 1$	$\leqslant 1$	1.2	2.13	1.84	1.75	a
[48]; $n = 0$, $R = CH_2CH_2OH$	1.0	$\leqslant 1$	1.24	4.06	3.79	3.23	a
[36], [2.2.1]-Cryptand	5.4	3.95	<2	6.95	7.35	6.30	b
[38]; $n = 1$, $R = CH_3$	—	1.1	1.0	2.38	4.17	3.78	a
[49]; $n = 1$, $R = CH_2CH_2OH$	—	$\leqslant 1$	$\leqslant 1$	3.71	4.30	5.33	a
[37], [2.2.2]-Cryptand	3.9	5.4	<2	4.40	8.00	9.50	b

[a] Gramain and Frère, 1979
[b] Lehn and Sauvage, 1971

compound [38]. A comparison with [2.2.1]-cryptand [36] and [2.2.2]-cryptand [37], having the same number of binding sites as [48] and [49] respectively, reveals that two CH_2CH_2OH units still have a much smaller stabilizing effect than one $CH_2CH_2OCH_2CH_2OCH_2CH_2$ bridge (Table 8).

The influence of additional functional groups on the complexing abilities of 1,3-xylyl-crown ethers [28] and [50]–[53] and binaphthyl-crown ethers

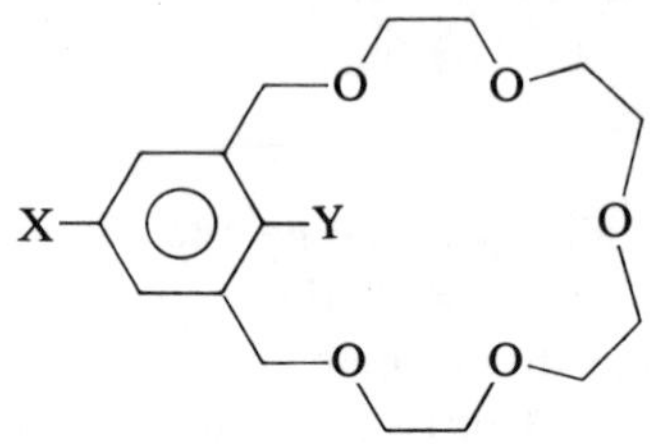

[28], $X = Y = H$

[50], $X = H, Y = CO_2CH_3$

[51], $X = CH_3, Y = OCH_3$

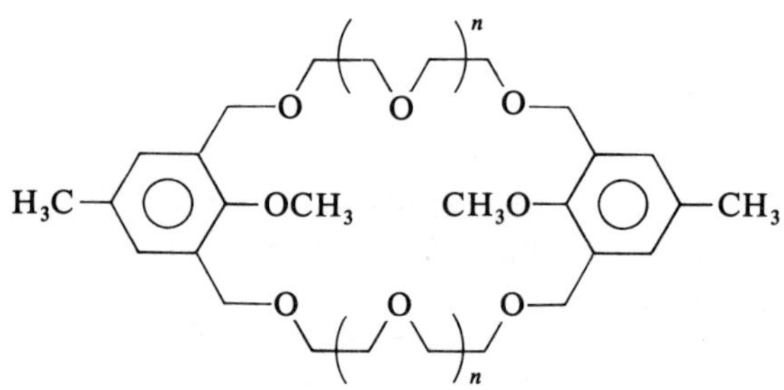

[52], $n = 0$

[53], $n = 1$

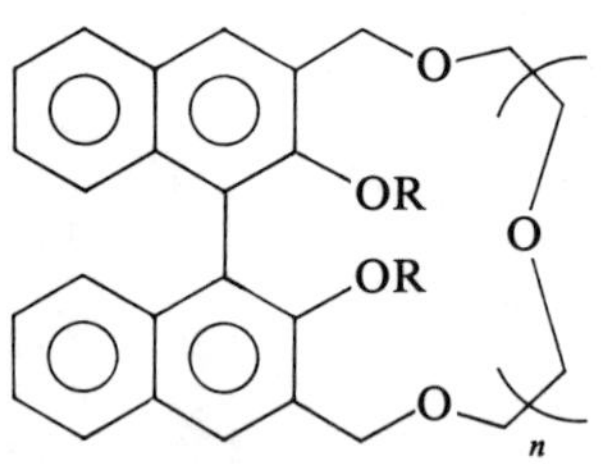

[54], $n = 2, R = CH_3$

[55], $n = 3, R = CH_3$

[56], $n = 2, R = H$

[57], $n = 3, R = H$

[54]–[57] has been studied by Koenig *et al.* (1976). The binding capacity for all the cations studied is markedly enhanced by the introduction of a binding site at the 2-position of 1,3-xylyl-18-crown-5 (Table 9). For all cations except K^+, the ester group is more effective than the methoxy group. Ligands [52] and [53], in which the OCH_3 groups can be placed above and below the cavity, show a poor cation binding capacity and a poor selectivity. Excessive flexibility of the ring was offered as an explanation. The binaphthyl series nicely

TABLE 9

Log K_{ip} values for picrate salt complexes in chloroform at 25°C[a]

Ligand	Li^+	Na^+	K^+	Cs^+	NH_4^+
[28]	3.08	3.38	4.90	4.58	4.87
[50]	4.30	5.95	6.09	5.68	5.70
[51]	4.00	4.67	6.26	5.06	5.56
[52]	2.93	3.30	3.59	3.28	3.28
[53]	2.75	3.70	3.94	3.78	3.93
[54]	4.21	6.09	6.57	5.75	5.90
[55]	4.11	4.63	5.98	5.98	5.68
[56]	3.34	3.90	4.61	4.40	3.77
[57]	3.85	4.19	5.37	5.30	4.85

[a] Koenig *et al.*, 1976

illustrates the difference between OCH_3 and OH binding sites. For Na^+, K^+, Cs^+ and NH_4^+ the stabilizing effect of methoxy is at least an order of magnitude larger than that of hydroxy. This result was explained in terms of intramolecular hydrogen bonding in the hydroxy ligands, an interaction which must be disrupted to allow the formation of a complex with the cation. For larger cavities ([55] and [56]) the difference between OCH_3 and OH decreases.

Additional macrocyclic polyether groups show no co-operative effect when the polyether rings diverge. For instance, the bis(crown ethers) [58] and [59]

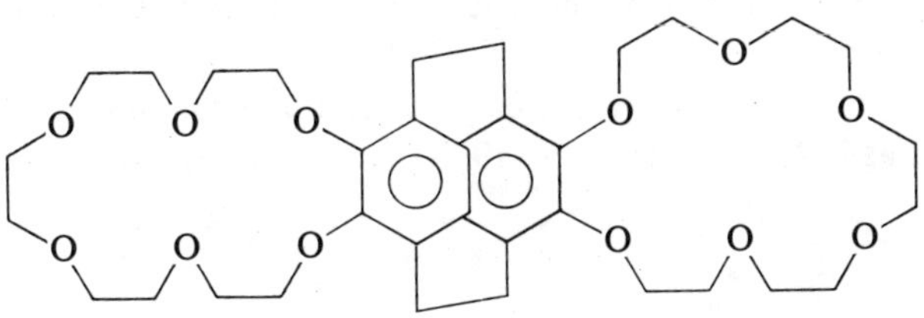

[58]

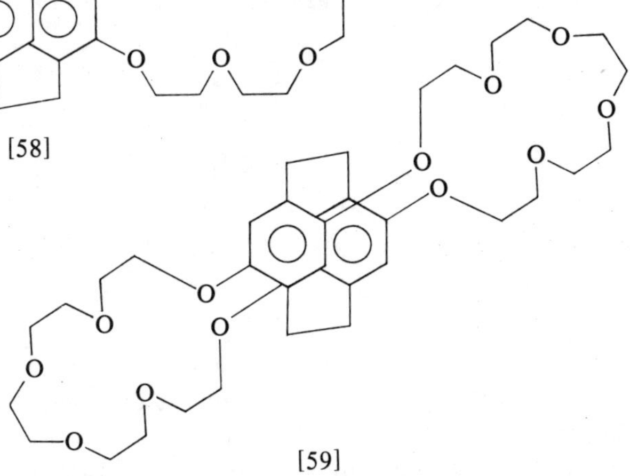

[59]

prepared by Cram's group did not show an enhanced binding capacity towards alkali cations (in the form of picrate salts in chloroform solution) as compared with 2,3-naphtho-18-crown-6 [60] (Helgeson *et al.*, 1977). Bis(crown ethers) having two benzo-15-crown-5 moieties in close proximity (such as [61]) have

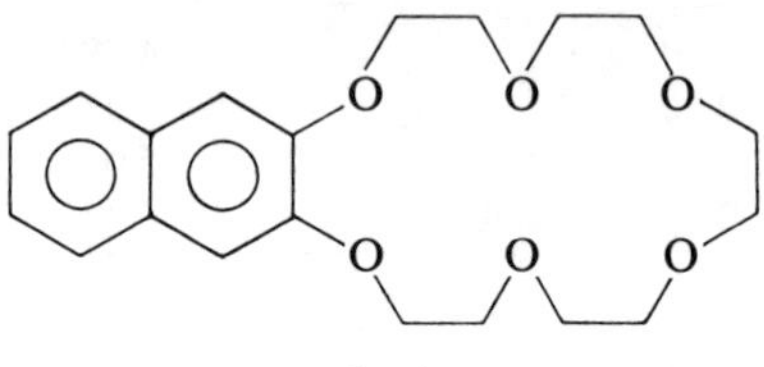

[60]

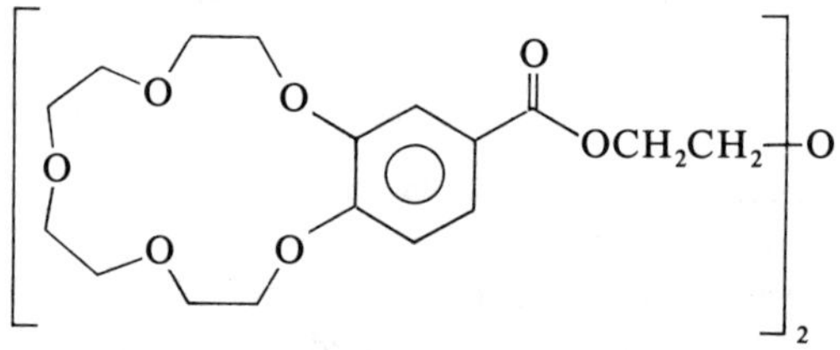

[61]

been studied by Smid's group (Wong *et al.*, 1974; Bourgoin *et al.*, 1975). The binding constant for the potassium picrate ion pair in tetrahydrofuran was at least one order of magnitude higher for [61] than for 15-crown-5 [2]. The synergistic effect of the two cavities was weakened on increasing the distance between them (Bourgoin *et al.*, 1975).

Co-operative effects of crown units are particularly pronounced in poly(crown ethers) such as [62] (Kimura *et al.*, 1978) and [63] (Kopolow *et al.*,

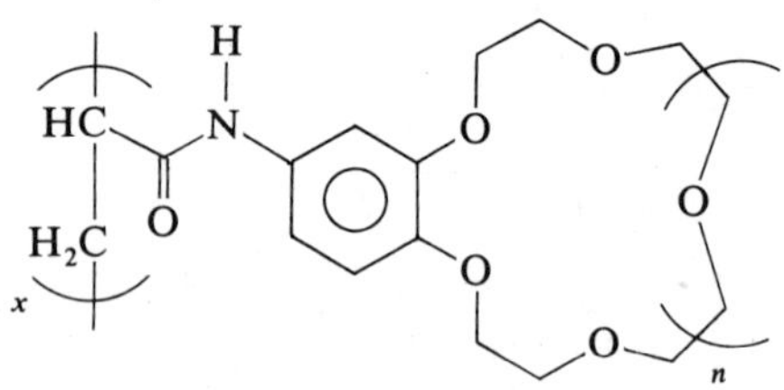

[62]

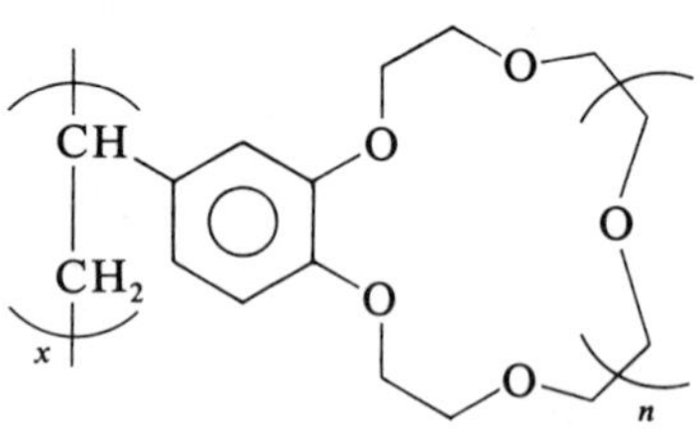

[63]

1973). Whereas the binding constants (expressed per crown unit) for small cations (such as Li^+, Na^+) are almost the same for polymer and monomer, larger cations (such as K^+, Rb^+, and Cs^+)—which tend to form 2:1 complexes with the monomer—are preferentially bound by the polymer.

Studies of cation binding by poly(crown ethers) have revealed very interesting features, but a detailed discussion of this subject would be beyond the scope of this review. For the pertinent literature the reader is referred to recent articles by Smid (1976), Wong and Smid (1977) and references cited therein.

Very stable complexes can be formed when there is a good match between cation and crown ether not only with respect to size but also with respect to charge. This is apparent from the work of Cram's group, who examined crown ethers with additional anionic binding sites. For example, they found the monovalent ligand [64] to form extremely stable 1 : 1 complexes with K^+. With Ba^{2+}, [64] forms very stable 2 : 1 complexes. The bivalent ligand [65] binds

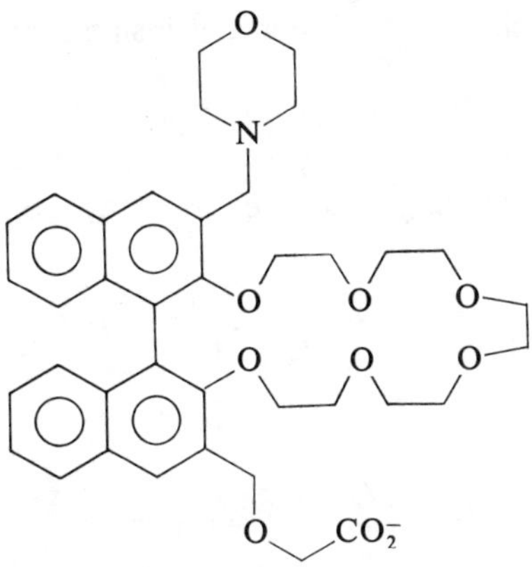

[64]

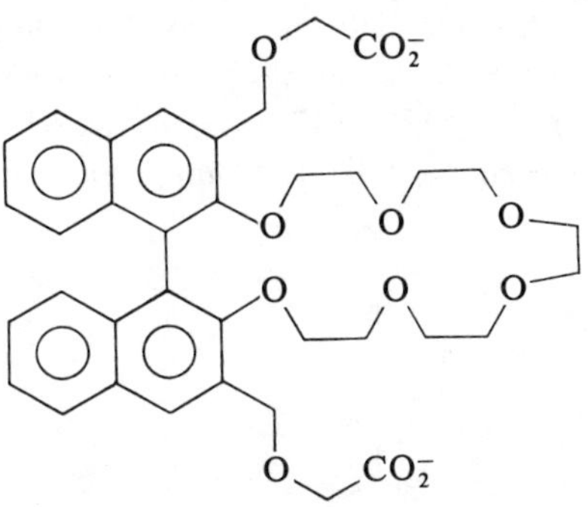

[65]

Sr^{2+} so well that it was selectively extracted from a bulk sample of a barium salt (Helgeson *et al.*, 1973a). Binding constants for metal-cation complexes of 1,3-xylyl-crown ethers [66]–[69] carrying an additional carboxylate binding

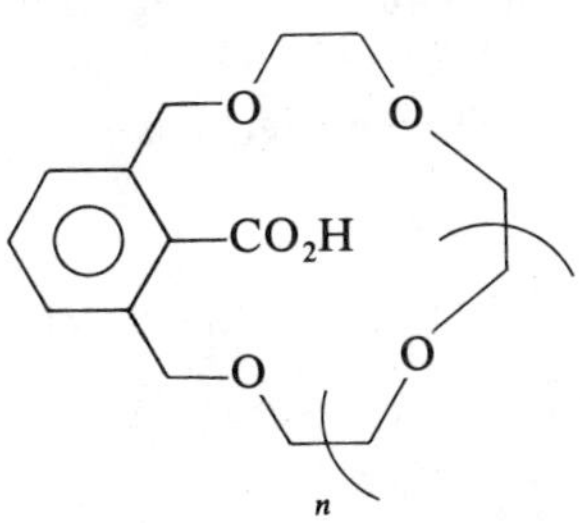

[66], $n = 1$

[67], $n = 2$

[68], $n = 3$

[69], $n = 6$

site have not been determined yet, but the anionic species readily extracted Li^+, Na^+, K^+ and Ca^{2+} from their aqueous solutions (Newcomb *et al.*, 1977a). The binding capacity of ligand [70] with bivalent cations in water has been studied by Takagi *et al.* (1978), who found the value of log K for Ca^{2+}, Sr^{2+}, and

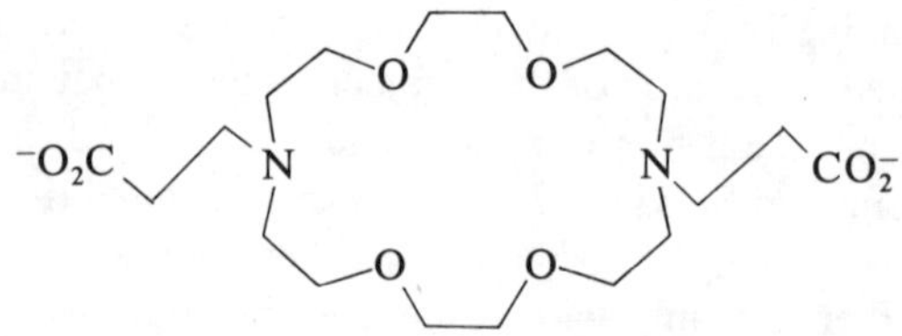

[70]

Ba^{2+} to be 4.0, 4.4, and 3.8 respectively. Comparison of these values with those for the parent compound [38] and ligand [49] having OH instead of CO_2^- groups (Table 8) reveals a disappointingly small effect of the carboxylate anions. In fact, it would seem that Ba^{2+} strongly prefers OH to CO_2^- as a binding site. It is clear, however, that the amount of evidence is insufficient to justify any generalization of this observation.

Anionic units have not only been attached to macrocyclic polyethers via flexible arms but have also been incorporated into the cycle itself. Alberts and Cram (1976, 1977) have studied the ion-binding properties of crown ethers containing β-diketone units, such as [72] and [73].

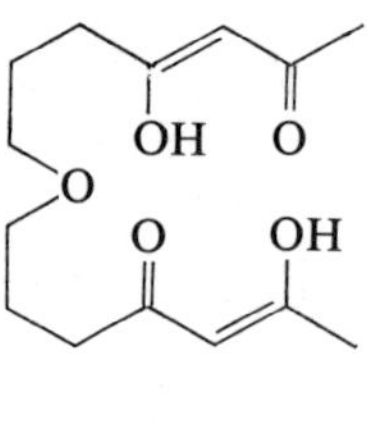

[71]

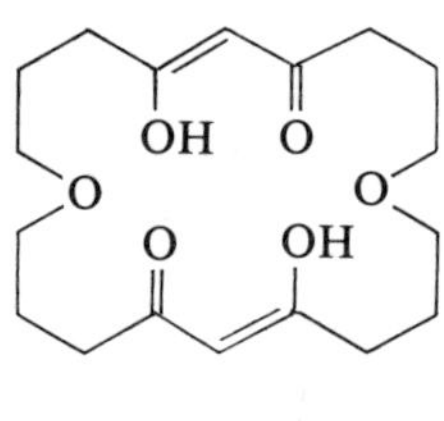

[72]

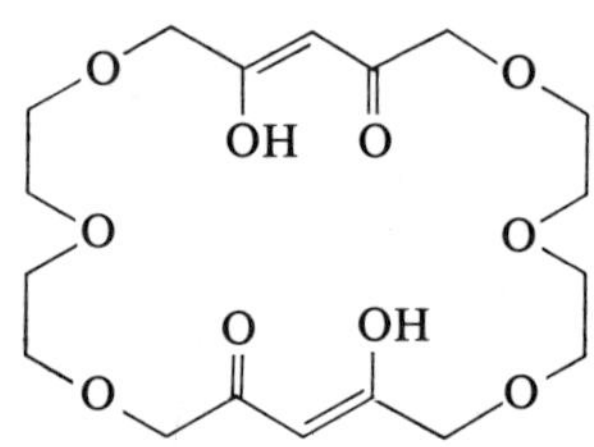

[73]

Pre-orientation of two acetylacetone (acac) units to form [71] increases the binding constants for various bivalent cations by 2.5–3.9 powers of ten (Table 10). The change in binding capacity caused by ring formation is indicated by the difference between [72] and [71]. For all the cations the macrocyclic effect is favourable, although the gain in binding energy is much smaller than for the

TABLE 10

Log K values for ligands [71]–[73] in water/dioxan (1/1) at 24°C[a]

Ligand	Cu^{2+}	Zn^{2+}	Pb^{2+}	Mg^{2+}	Ca^{2+}	Sr^{2+}	Ba^{2+}
Acetylacetone	9.5	6.1	5.5	4.5	3.1	2.7	2.5
[71]	12.0	8.8	8.6	7.1	7.0	5.3	5.2
[72]	12.1	9.5	9.1	7.2	7.4	5.6	5.4
[73]	11.3	9.7	—	7.4	7.8	6.7	6.4

[a] Alberts and Cram, 1976, 1977

gathering of two acac units. On increasing the size of the cavity to that of [73] the binding capacity for Sr^{2+} and Ba^{2+} increases.

ELECTRON DENSITY AT THE BINDING SITES

Replacing a CH_2CH_2-unit in 18-crown-6 by a benzo-unit reduces the electron-density at two of the oxygen binding sites. The data in Tables 3 and 4 show that this effect results in a marked drop in binding capacity. Crown ethers containing only oxygen atoms conjugated with benzene rings, such as hexabenzo-18-crown-6 [74] (Kime and Norymberski, 1977) and tri-binaphthyl-24-crown-6 [75] (De Jong *et al.*, 1975) showed no sign of complexation.

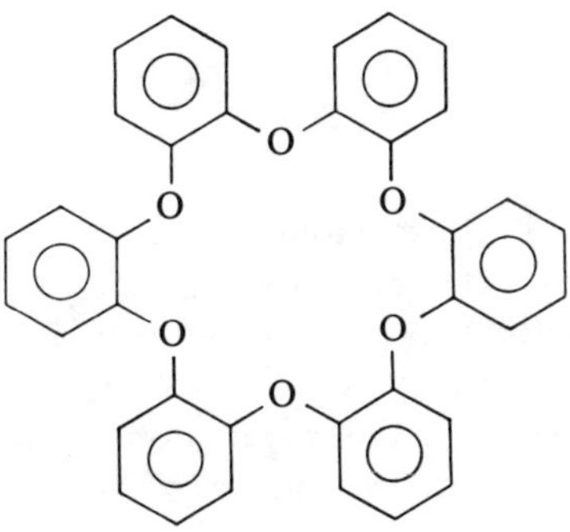

[74]

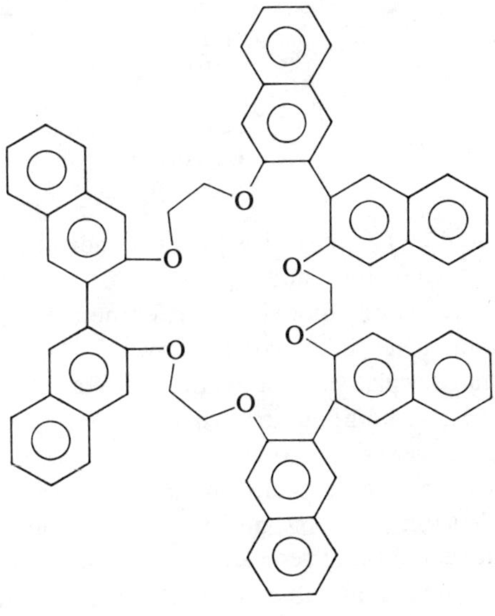

[75]

The effect of substituents on the ion-binding capacity of benzo-crown ethers [76] has been studied in acetone solution (Ungaro *et al.*, 1976). For the

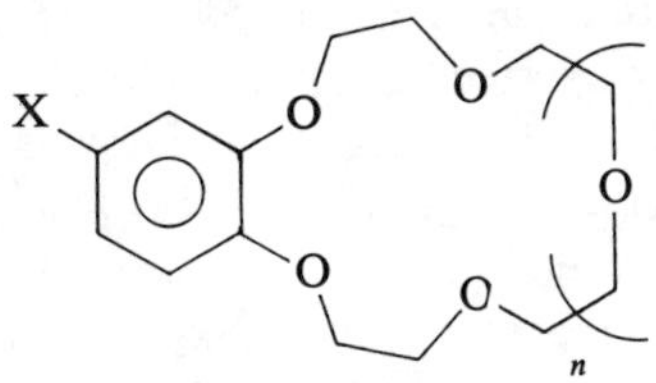

[76]

complexes of Na^+ with benzo-15-crown-5 ligands (Table 11) a good Hammett correlation was obtained by plotting log K_i versus $(\sigma_p^- + \sigma_m)$, the ρ-value being

TABLE 11

Log K_1 values for substituted benzo-crown ethers [76] in acetone at 25°C[a]

X	$n = 1$ Na^+	$n = 2$ Na^+	K^+
NH_2	3.91	—	—
CH_3	3.60	5.09	5.58
H	3.54	4.72	5.10
Br	3.31	—	—
COOH	3.21	—	—
CO_2CH_3	3.09	—	—
CHO	3.05	4.59	4.89
$CONHC_4H_9$	—	4.51	4.75
NO_2	2.56	4.67	4.80

[a] Ungaro *et al.*, 1976

−0.45. The negative ρ-value, indicating that electron-attracting substituents decrease the binding capacity, is in line with expectations based on the electrostatic model. No Hammett correlations were found for Na^+ and K^+ complexes of benzo-18-crown-6. The observation that benzo-15-crown-5 is more sensitive to substituent effects than benzo-18-crown-6 was explained by the last compound having four instead of three aliphatic oxygen atoms. The absence of a good Hammett correlation for Na^+ complexes of benzo-18-crown-6 was attributed to the cavity being too large for the cation. For electron-releasing and electron-attracting substituents the cation might be attracted to, or repelled by, the benzo-moiety (Ungaro *et al.*, 1976).

Significant substituent effects have also been reported for Na^+ complexes of 4,4′-substituted dibenzo-18-crown-6 in dimethylformamide (Shchori and Jagur-Grodzinski, 1973; Shchori *et al.*, 1973). The extraction of sodium and potassium 4-(*p*-dimethylaminophenylazo)-benzenesulfonates from water into methylene chloride by substituted dibenzo-18-crown-6 has been studied by Pannell *et al.* (1977). A plot of log K_{ext} against $\Sigma\sigma$ (using σ_m, σ_p, and σ_o) gave a reasonably good, although not perfect, Hammett correlation, with ρ values for Na^+ and K^+ of −1.44 and −3.35, respectively. It was also found that the K^+/Na^+ selectivity was reversed on going from electron-releasing to electron-attracting substituents. A decrease in K^+/Na^+ selectivity with destabilizing substituents has also been reported for 1,3-xylyl-18-crown-5 compounds [77] in chloroform solution (Moore *et al.*, 1977). Relative to Y = H, CN destabilizes K^+ complexes, shows little effect on Na^+ binding, and even increases the ability to bind Li^+ (Table 12). The striking differences in response

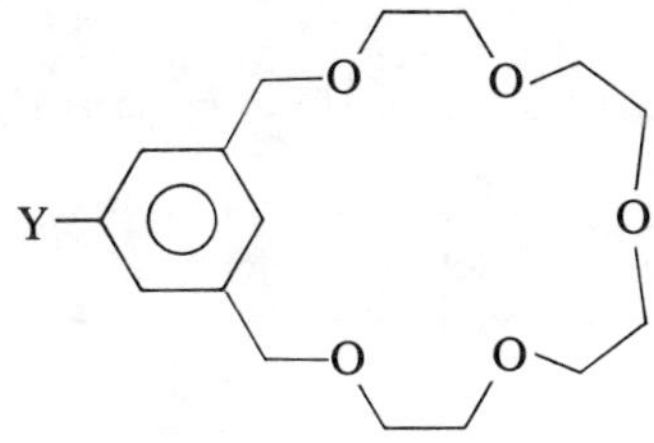

[77]

TABLE 12

Log K_{ip} values for picrate salt complexes with crown ethers [77] in $CDCl_3$ at 24°C[a]

Y	Li^+	Na^+	K^+	Rb^+	Cs^+	NH_4^+
H	2.00	3.23	5.04	4.82	4.70	5.06
Br	2.00	3.04	4.56	4.28	4.16	4.60
$C(CH_3)_3$	2.40	3.20	5.04	4.85	4.70	5.08
OCH_3	2.60	3.41	5.12	4.87	4.74	5.09
CN	3.00	3.20	4.08	3.74	3.52	4.03

[a] Moore *et al.*, 1977

to substituent effects between Li^+ and Na^+, on the one hand, and K^+, Rb^+, Cs^+ and NH_4^+ on the other, were attributed to the fact that the small cations can minimize substituent effects by selecting the most basic sites, whereas cations that are too large have to contact all binding sites. Although the binding constants showed a poor correlation with existing σ-constants, a good linear free energy correlation was observed when K^+, Rb^+, and Cs^+ picrates were compared with NH_4^+ picrate as a standard. The value of 1.03 obtained for the slope indicates that cations which fill the cavity of compounds [77] respond in the same way to changes in π-basicity of the aromatic nucleus.

The effect of carbonyl adjacent to oxygen binding sites has been studied for compounds [3] and [78] (Izatt *et al.*, 1977a, 1977b) and the pyridino-crown

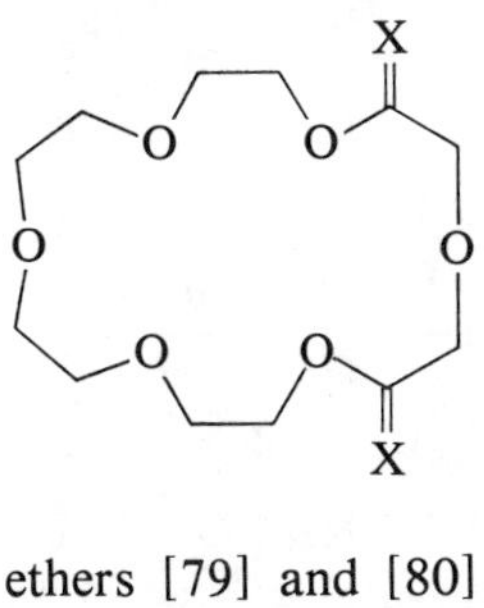

[3], X = H_2

[78], X = O

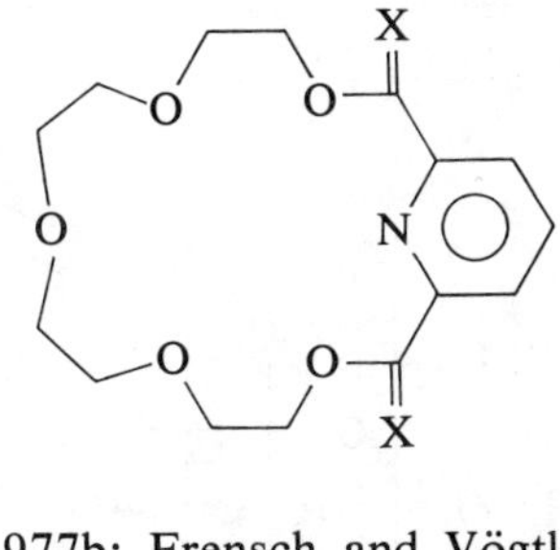

[79], X = H_2

[80], X = O

ethers [79] and [80] (Izatt *et al.*, 1977b; Frensch and Vögtle, 1977). The introduction of the carbonyl group into 18-crown-6 lowered the binding

TABLE 13

Log K, $-\Delta H$ and $T\Delta S$ values for complexation of crown ethers [3], [78]–[80] in methanol at 25 °C[a]

	Log K			$-\Delta H$ (kcal mol^{-1})			$T\Delta S$ (kcal mol^{-1})		
Ligand	Na^+	K^+	Ba^{2+}	Na^+	K^+	Ba^{2+}	Na^+	K^+	Ba^{2+}
18-Crown-6 [3]	4.36	6.05	7.0	8.4	13.4	10.2	−2.4	−5.2	0.7
[78]	2.5	2.79	3.1	2.27	5.87	0.46	1.1	−2.1	3.8
[79]	4.09	5.35	>6	5.44	9.11	7.72	0.14	−1.8	0.5
[80]	4.29	4.66	4.34	6.19	9.3	6.03	−0.34	−2.9	−0.11

[a] Izatt *et al.*, 1977a, b

constants of Na^+, K^+, and Ba^{2+} by a factor of 10^2–10^4 (Table 13). The unfavourable change in free energy is entirely due to the enthalpy of binding. The entropy contribution becomes even more favourable. Quite a different behaviour is observed with the pyridino-crown ethers: the addition of two carbonyl groups to [79] lowers the binding capacity for K^+, and Ba^{2+}, but increases that for Na^+. While the entropy term favours complexes of the carbonyl compound, the enthalpy term varies with no clear trend. It was suggested that in the case of the pyridino-crown ether, the aromatic nucleus might also contribute to the binding energy. The carbonyl groups did not seem to be directly involved in cation binding (Izatt *et al.*, 1977b).

NITROGEN AND SULFUR DONOR SITES

Several investigators have studied the effect of substituting N or S for O on the cation affinity of crown ethers. Log K values for 15-crown-5 [2] and

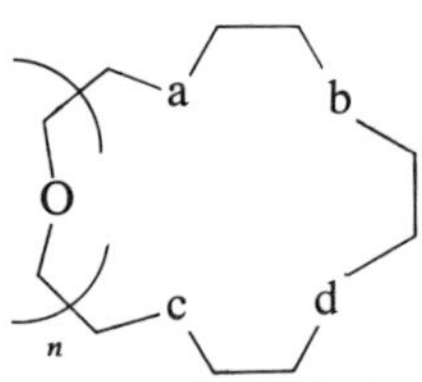

	n	a	b	c	d
[2]	1	O	O	O	O
[81]	1	O	O	O	S
[82]	1	O	NH	O	NH
[83]	1	NH	S	S	NH
[3]	2	O	O	O	O

	n	a	b	c	d
[84]	2	O	O	O	NH
[85]	2	O	O	O	S
[86]	2	NH	O	NH	O
[87]	2	S	O	S	O
[88]	2	O	O	S	S

derivatives [81]–[83], and for 18-crown-6 [3] and derivatives [84]–[88] are collected in Table 14. In K^+, Tl^+, and Ba^{2+} complexes, substitution of N or S for O causes a considerable drop in binding constants. The complex stabilities

TABLE 14

Effect of N and S substitution on log K for 1:1 complexes in water (W) or methanol (M) at 25°C

Ligand	K^+ W	K^+ M	Tl^+ W	Ag^+ W	Ba^{2+} W	Pb^{2+} W
15-Crown-5 [2]	0.74[a]	—	1.23[a]	0.94[a]	—	1.85[b]
[81]	—	—	0.8[b]	5.0[b]	—	1.65[b]
[82]	1.0[d]	—	—	5.85[e]	1.0[d]	5.85[e]
[83]	—	—	—	8.95[e]	—	5.67[e]
18-Crown-6 [3]	2.03[a]	6.10[c]	2.27[a]	1.50[a]	3.78[a]	4.27[b]
[84]	—	3.90[c]	—	3.3[c]	—	—
[85]	—	3.61[b]	—	—	—	—
[86]	<1[d]	2.04[c]	1.1[f]	7.8[c]	2.51[d]	6.9[e]
[87]	—	1.15[c]	0.93[b]	4.34[c]	—	3.13[b]
[88]	—	—	1.38[b]	3.0[b]	—	2.63[b]

[a] Izatt *et al.*, 1976a
[b] Izatt *et al.*, 1978a
[c] Frensdorf, 1971a
[d] Gramain and Frère, 1979
[e] Arnaud-Neu *et al.*, 1977
[f] Anderegg, 1975

decrease in the same order as the electronegativity of the heteroatoms: O > NH > S. This is not unexpected since cation–dipole interactions are thought to provide the binding energy of crown-ether complexes with hard cations. Yet the magnitude of the effect of N substitution seems rather large (for example, a difference of 3 kcal mol^{-1} between the free energies of binding for K^+ complexes of 18-crown-6 [3] and [2.2]-cryptand [86]), considering that in the gas phase the potassium affinities of nitrogen and oxygen bases show only very small differences (Davidson and Kebarle, 1976). Therefore, other factors, such as conformational effects and specific solvent–ligand interactions, might also contribute to the lower stability of complexes of aza-crown ethers.

For soft cations, such as Ag^+ and Pb^{2+}, covalent contributions are much more important, and consequently the observed order of complex stabilities is quite different from that for alkali cations: NH > O > S for Pb^{2+} and NH, S > O for Ag^+. Dissection of the overall effect into enthalpy and entropy contributions (Table 15) reveals the complicated nature of the heteroatom effect. For K^+ and Ba^{2+}, the more favourable entropy contribution for N and S ligands is more than offset by the unfavourable change in enthalpy of binding.

TABLE 15

Changes in enthalpy and entropy on replacing oxygen by nitrogen and sulfur in $X(CH_2CH_2OCH_2CH_2OCH_2CH_2)_2Y$

Cation	Solvent	X	Y	$-\Delta\Delta H^a$	$T\Delta\Delta S^a$	Ref.
K^+	Methanol	O	S	−5.4	+1.1	b
Tl^+	Water	S	S	+6.6	−8.4	b
Ag^+	Water	N	N	+7.0	+1.6	c
		S	S	+14.5	−10.7	b
Ba^{2+}	Water	N	N	−4.6	+3.4	c
Pb^{2+}	Water	S	S	+16.0	−17.6	b

[a] In kcal mol^{-1}; $\Delta\Delta H = \Delta H$ (compound) − ΔH (18-crown-6)
$\Delta\Delta S = \Delta S$ (compound) − ΔS (18-crown-6)

[b] Izatt *et al.*, 1978a, 1977a

[c] Anderegg, 1975

For Tl^+ and Pb^{2+}, the S ligands are much more stable with regard to the enthalpy, which difference is more than offset by the unfavourable change in entropy. For Ag^+, N ligands are preferred to O ligands both from an enthalpy and from an entropy point of view. A detailed explanation of this rather different behaviour is not yet available.

The cation affinity of aza-crown ethers depends on the type of substituent attached to the nitrogen. Wester and Vögtle (1978) have determined the cation binding constants for substituted [2.2]-cryptands [38] and [89]–[91] in

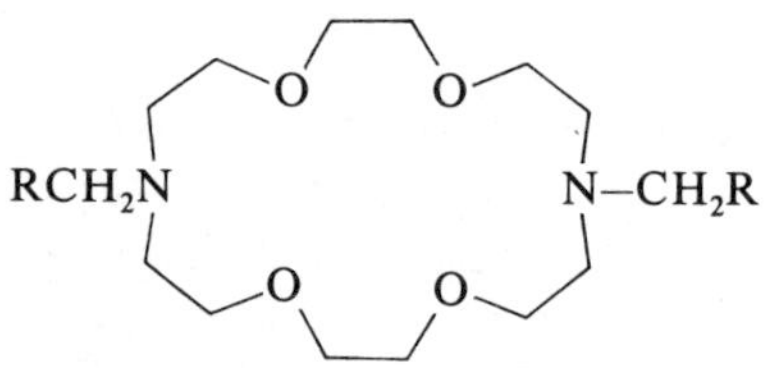

[38], R = H

[89], R = t-Bu

[90], R = C_6H_5

[91], R = *p*-$CH_3OC_6H_4$

methanol/water (45/6). Complexes with Na^+ showed almost the same log *K* values for [38], [89], [90], and [91]: viz. 2.55, 2.2, 2.4, and 2.2 respectively. Benzyl-substituted compounds showed a considerably weaker affinity for Sr^{2+} and Ba^{2+}; log *K* (Ba^{2+}) was 5.95 for [38] and 4.5 for both [90] and [91]. A drastic drop in binding constants was observed with the neopentyl substituent [log *K* (Ba^{2+}) = 2.4]. This effect seems stereochemical rather than electronic in nature.

SOLVENT EFFECTS

Both the absolute and the relative values of the association constants of crown-ether complexes with cations are dependent on the type of solvent used.

$$\begin{array}{ccc} M^+_{aq} + Cr_{aq} & \overset{K_W}{\rightleftharpoons} & (M^+.Cr)_{aq} \\ \downarrow \quad \downarrow & & \downarrow \\ M^+_S + Cr_S & \underset{K_S}{\rightleftharpoons} & (M^+.Cr)_S \end{array} \qquad (6)$$

The important parameters for substituting a solvent S for water (W) follow from the thermodynamic cycle (6). K_S and K_W are related through the free energy of transfer terms (7). The enthalpy of transfer is given by (8).

$$-RT \ln (K_S/K_W) = \Delta\Delta G = \Delta G_{tr} (M^+.Cr) - \Delta G_{tr} (Cr) - \Delta G_{tr} (M^+) \qquad (7)$$

$$\Delta\Delta H = \Delta H_{tr} (M^+.Cr) - \Delta H_{tr} (Cr) - \Delta H_{tr} (M^+) \qquad (8)$$

Effects of solvents on binding constants are frequently discussed in terms of differences in cation solvation. It is clear from (7), however, that in addition to the free energies of transfer for the cations, the free energies for the complexed cation and the ligand must be considered. The importance of ligand solvation was nicely demonstrated by Abraham *et al.* (1977) for [2.2.2]-cryptand. The overall transfer of the Na^+ complex from water to methanol was exothermic ($\Delta\Delta H = -3.4$ kcal mol^{-1}), the relative contributions being: $\Delta H_{tr} (M^+.Cr) = 5.6$, $\Delta H_{tr} (Cr) = 13.9$, and $\Delta H_{tr} (M^+) = -4.9$. It is not known yet whether crown ethers show similar large effects of ligand solvation.

Gutknecht *et al.* (1978) have found that $\Delta\Delta G + \Delta G_{tr} (M^+)$ is roughly constant (-5.1 ± 0.6 kcal mol^{-1}) for K^+, Ag^+, and Tl^+ complexes of [2.2.2]-cryptand on transfer from water to non-aqueous solvents (Table 16). The value for methanol (-4.2 kcal mol^{-1}) calculated from the log K values determined by Cox *et al.* (1978) is in the same range. Consequently, $\Delta G_{tr} (M^+.Cr) - \Delta G_{tr} (Cr)$ is very small on transfer between aprotic solvents, and therefore differences in binding constants and selectivities among apolar solvents are mainly due to differences in the free energy of transfer for the cation. Since solvent donor numbers (Gutmann, 1968) correlate reasonably well with $\Delta G_{tr} (M^+)$ values, donor numbers and binding constants are also correlated in this particular case. This relationship was also observed for the effect of solvent on the cation exchange rates of cryptates (Popov, 1976).

The thermodynamic data for K^+ complexes of dibenzo-18-crown-6 [11] in aprotic solution show that $\Delta\Delta G + \Delta G_{tr} (K^+)$ changes with solvent (Table 16). Although the data have been collected from several sources, they would seem to indicate that the cation is more exposed to interactions with the solvent in crown-ether complexes than in cryptates. Hence, the term

$$\Delta G_{tr} (M^+.Cr) - \Delta G_{tr} (Cr)$$

TABLE 16

Free energies of transfer[a] of K^+ complexes of dibenzo-18-crown-6 [11] and [2.2.2]-cryptand [37] from water to non-aqueous solvents at 25°C

Solvent	Donor no.[b]	$\Delta G_{tr}(K^+)$[c]	$\Delta\Delta G$ [11][d]	$\Delta\Delta G$ [37]
Methanol	25.7	2.4	−4.6	−6.6[f]
Dimethyl sulfoxide	29.8	−2.9	−2.5	−2.0[e]
Dimethylformamide	26.6	−2.3	−2.7	−3.4[e]
Propylene carbonate	15.1	1.4	−4.7	−7.6[e]
Acetonitrile	14.1	1.9	−4.4	−7.2[e]

[a] In kcal mol^{-1}
[b] Gutmann, 1968
[c] Cox *et al.*, 1974
[d] Data taken from Table 17
[e] Gutknecht *et al.*, 1978
[f] Cox *et al.*, 1978

is expected to be solvent-dependent for crown ethers. This effect is not so large that it will change the order of relative complex stabilities for K^+: propylene carbonate > acetonitrile > dimethylformamide > dimethyl sulfoxide > water, for both 18-crown-6 and [2.2.2]-cryptand.

Log K values for various complexes of 18-crown-6 [3], benzo-18-crown-6 [9], and dibenzo-18-crown-6 [11] in several solvents are collected in Table 17. For Na^+ and K^+, the binding constants roughly correlate with solvent donor numbers, although some irregularity does exist. The binding constants for the larger cations (Rb^+, Cs^+) show remarkably little dependence on the type of aprotic solvent. In protic solvents, the binding constants were found to increase regularly for all alkali cations in the order: water < methanol < ethanol < n-propanol (Agostiano *et al.*, 1976).

The sequence of the selectivities towards cations is also solvent dependent: for dibenzo-18-crown-6 [11] the sequence is $K^+ > Na^+ > Rb^+ > Cs^+$ in water, methanol, dimethylformamide and dimethyl sulfoxide (Dechter and Zink, 1976; Srivanavit *et al.*, 1977), whereas it is $Na^+ > K^+ > Rb^+ > Cs^+$ in acetonitrile (Hofmanova *et al.*, 1978). A reversal of the K^+/Na^+ selectivity on going to apolar aprotic solvents was also observed for fluorenyl salts (Wong *et al.*, 1970). Whereas for alkali cations the sequence of binding constants and enthalpies are the same in water (Izatt *et al.*, 1976a), they differ considerably in methanol/water mixtures (Izatt *et al.*, 1976b), dimethyl sulfoxide and acetone (Arnett and Moriarity, 1971).

The cation selectivity in extraction experiments is dependent on differences in both the distribution constants K_D and the binding constants K_{lp} in the water-saturated organic solvent (3). The extractability of alkali cations into

TABLE 17

Effect of solvent on log K

Crown ether	Solvent	Na^+	K^+	Rb^+	Cs^+	Tl^+	Ba^{2+}
18-crown-6 [3]	Water	0.80[a]	2.03[a]	1.56[a]	0.99[a]	2.27[a]	3.78[a]
	Water/methanol (70/30)	2.76[i]	4.33[i]	3.46[i]	2.84[i]	—	6.0[i]
	Methanol	4.32[e]	6.10[e]	5.35[f]	4.62[e]	—	7.0[g]
	Dimethyl sulfoxide	—	—	—	3.04[r]	—	—
	Dimethylformamide	—	—	—	3.95[r]	—	—
	Propylene carbonate	—	—	—	4.17[r]	—	—
Benzo-18-crown-6 [9]	Water	—	2.0[b]	—	—	—	3.7[b]
	Methanol	—	5.20[h]	—	—	4.60[h]	—
	Acetone	4.72[l]	5.1[l]	—	—	—	—
	Acetonitrile	4.90[h]	5.30[h]	4.40[h]	4.05[h]	5.70[h]	—
Dibenzo-18-crown-6 [11]	Water	1.1[c]	1.6[c]	1.1[c]	0.83[d]	1.5[d]	1.95[d]
	Methanol	4.36[e]	5.00[e]	—	3.55[e]	4.00[h]	—
	Dimethyl sulfoxide	3.31[j]	3.43[j]	3.37[j]	3.31[j]	—	—
	Dimethylformamide	3.34[j]	3.55[j]	3.54[j]	3.49[j]	2.45[o]	—
		2.78[q]	—	—	—	—	—
	Propylene carbonate	3.87[j]	5.08[j]	3.76[j]	3.52[j]	—	—
	Dimethoxyethane	3.73[k]	—	—	—	—	—
	Acetonitrile	5.01[m]	4.80[m]	—	—	4.11[n]	—
		5.00[h]	4.70[h]	3.70[h]	3.50[h]	4.90[h]	—
	Acetone	4.6[b]	—	—	>3[r]	3.3[n]	—
	Tetrahydrofuran	4.6[p]	—	—	—	—	—

[a] Izatt *et al.*, 1976a
[b] Live and Chan, 1976
[c] Harris, 1977
[d] Shchori *et al.*, 1975
[e] Frensdorf, 1971a
[f] Burden *et al.*, 1977
[g] Izatt *et al.*, 1977a
[h] Hofmanova *et al.*, 1978
[i] Izatt *et al.*, 1976b
[j] Matsuura *et al.*, 1976
[k] Shchori and Jagur-Grodzinski, 1973
[l] Ungaro *et al.*, 1976
[m] Evans *et al.*, 1972
[n] Platzner *et al.*, 1978
[o] Srivanavit *et al.*, 1977
[p] Wong *et al.*, 1974
[q] Shchori *et al.*, 1973
[r] Mei *et al.*, 1977b

organic solvents usually follows the order: $Cs^+ > Rb^+ > K^+ > Na^+ > Li^+$ (Haynes and Pressman, 1974). The observed overall selectivity in extractions of picrates into dichloromethane (Jaber *et al.*, 1977; Jawaid and Ingman, 1978; Frensdorf, 1971b) and benzene (Iwachido *et al.*, 1978; Sekine *et al.*, 1978; Sadakane *et al.*, 1975) was $K^+ > Tl^+ > Rb^+ > Cs^+ > Na^+ > Li^+$ for 18-crown-6 [3], dibenzo-18-crown-6 [11] and dicyclohexyl-18-crown-6 ([20] and [21]). The potassium picrate extractability decreased in the order dichloromethane > benzene > n-hexane.

From the data available so far it is clear that basic patterns in the solvent effect start to become recognizable, but that further studies are needed in order to arrive at a model that has any predictive value.

ANION EFFECTS

As outlined in the introduction of this section, a crown ether can form complexes with both the free cation (measured as K_i) and the ion pair (measured as K_{ip}). With a view to studying the interaction between cation and crown ether, the effect of anions has deliberately been minimized either by using high dilution conditions or by choosing anions (such as tetraphenylborate) whose salts are predominantly dissociated in solution. Consequently, the effect of the type of anion on K_{ip} is still obscure. Shchori *et al.* (1975) have found that dibenzo-18-crown-6 [11] forms stronger complexes with $[BaCl]^+$ (log $K = 2.15$) than with Ba^{2+} (log $K = 1.95$). The authors attributed this effect to the favourable effect of (partial) charge neutralization. The same ligand showed no particular preference for either Na^+ or the ion pair $Na^+.[B(C_6H_5)_4]^-$ in dimethoxyethane solution (Shchori and Jagur-Grodzinski, 1973). In contrast, bis(crown ethers) such as [61] showed a strong preference for the free K^+ and NH_4^+ ions over the picrate ion pairs (K_i/K_{ip} ≈ 400; Bourgoin *et al.*, 1975). Specific anion–ligand interactions may occur when the ligand contains H-bond donor groups. NMR studies of K^+ complexes of [2.2]-cryptand [86] in acetonitrile indicated that the acetate anion formed hydrogen bonds with the NH groups of the ligand (Knöchel *et al.*, 1977b).

The anion effect on the extractability of salts from aqueous into organic solution (K_{ext}) has been studied by several workers. It was found that the extractability increased with increasing hydrophobicity of the anion (Jawaid and Ingman, 1978).

3 Kinetics of metal cation complexation

The results of a temperature jump relaxation study of the complexation of metal cations by dibenzo-30-crown-10 [14] in methanol led Chock (1972) to propose a two-step mechanism. The first step (9) comprises a rapid equilibrium

$$\mathrm{Cr'} \underset{k_{21}}{\overset{k_{12}}{\rightleftharpoons}} \mathrm{Cr''} \tag{9}$$

$$\mathrm{Cr''} + \mathrm{M^+} \underset{k_{32}}{\overset{k_{23}}{\rightleftharpoons}} \mathrm{Cr.M^+} \tag{10}$$

of two different conformations of the crown ether. In the second step (10) only one of these—arbitrarily taken as Cr″—is able to form a complex with the metal cation. This mechanism was fully substantiated by Eyring and co-workers. Using an ultrasonic absorption method, the position of the conformational equilibrium and the activation parameters for the first step (9) were established for 18-crown-6 [3] in aqueous solution (Liesegang *et al.*, 1978). The results (Table 18) show that K_{21} $(= k_{21}/k_{12})$ is much smaller than unity, so that the more abundant conformational isomer binds the metal cation. The

TABLE 18

Parameters[a] for equations (9) and (10) in which Cr≡18-crown-6 and $M^+ \equiv K^+$ in water at 25 °C[b]

$K_{21} = 1.6 \times 10^{-2}$	$k_{21} = (1.0 \pm 0.3) \times 10^7$	$k_{12} = (6.2 \pm 0.2) \times 10^8$
$\Delta H_0 = -2.6 \pm 1.0$	$\Delta H^{\ddagger}_{12} = 7.4 \pm 1.0$	$\Delta H^{\ddagger}_{21} = 10.0 \pm 1.0$
$\Delta S_0 = 0 \pm 0.2$	$\Delta S^{\ddagger}_{12} = 7.7 \pm 0.2$	$\Delta S^{\ddagger}_{21} = 7.7 \pm 0.2$
$K_{23} = 116$	$k_{23} = 4.3 \times 10^8$	$k_{32} = 3.7 \times 10^6$
$\Delta H_0 = -6.2$	$\Delta H^{\ddagger}_{23} = 4.0$	$\Delta H^{\ddagger}_{32} = 10.2$
$\Delta S_0 = -11.1$	$\Delta S^{\ddagger}_{23} = -8.0$	$\Delta S^{\ddagger}_{32} = 3.1$

[a] Units: k_{32}, k_{21}, k_{12}: s^{-1}; ΔH: kcal mol^{-1}; ΔS: cal mol^{-1} K^{-1}; k_{23}: M^{-1} s^{-1}; K_{23}: M^{-1}

[b] Liesegang *et al.*, 1978

activation enthalpies are $\Delta H^{\ddagger}_{21}$ 10.2 ± 1.0 and $\Delta H^{\ddagger}_{12} = 7.4 \pm 1.0$ kcal mol^{-1}. These authors also determined the characteristics for the complex forming step (10). With K^+, complex formation is a very rapid process ($k_{23} = 4.3 \times 10^8$ M^{-1} s^{-1} at 25 °C), but not diffusion controlled. The mechanistic pathway leading to the complex is generally described by a stepwise replacement of an indeterminate number of solvent molecules from the first solvation sphere of the cation by the binding sites of the crown ether (Liesegang *et al.*, 1978). The fact that the enthalpy of activation for decomplexation ($\Delta H^{\ddagger}_{32} = 10.2$ kcal mol^{-1}) is similar to the enthalpy of activation for the conformational change Cr″ → Cr′ led the authors to suggest that the energy barrier to decomplexation might simply be the energy required to effect a conformational rearrangement of the complexed ligand.

Table 19 contains rate and equilibrium data for cation complexes of 18-crown-6 [3] and 15-crown-5 [2]. The k_{23} values for alkali and alkaline-earth metal cations are the same with these ligands, which suggests that loss of cation solvation is the dominant factor in complex formation, cavity size being unimportant. For Tl^+, Ag^+, and Hg^{2+}, a small decrease in k_{23} is observed on going from 18-crown-6 to 15-crown-5 (Rodriguez *et al.*, 1978). Rates of decomplexation (k_{32}) are much higher for 15-crown-5 than for 18-crown-6. Furthermore, k_{32} values span a range of 2000 for 18-crown-6 and only 164 for 15-crown-5. These facts indicate, as the authors suggest, that the differences in overall binding constants can be accounted for by variations in the rate of decomplexation (Rodriguez *et al.*, 1977). Combining this result with the suggestion that activation enthalpies for decomplexation are mainly deter-

TABLE 19

Rate and equilibrium constants for metal-cation complexes of 15-crown-5 and 18-crown-6 in water at 25 °C

	18-crown-6 [3][a,b,d]			15-crown-5 [2][c,d]		
Cation	$k_{23} \times 10^{-8}$ /M^{-1} s^{-1}	$k_{32} \times 10^{-5}$ /s^{-1}	K_{23}/M^{-1}	$k_{23} \times 10^{-8}$ /M^{-1} s^{-1}	$k_{32} \times 10^{-5}$ /s^{-1}	K_{23}/M^{-1}
Li^+	~0.8	~600	~1.3	—	—	—
Na^+	2.2	340	6.3	2.4	480	4.7
K^+	4.3	37	115	4.3	780	5.5
Rb^+	4.4	120	36.3	4.6	1200	3.8
Cs^+	4.3	440	9.8	—	—	—
NH_4^+	5.6	440	12.6	—	—	—
Ag^+	11.2	350	31.6	6.4	740	8.7
Tl^+	9.0	48	1.86×10^2	7.1	420	17.0
Sr^{2+}	0.77	1.5	5.25×10^2	0.65	7.3	89.1
Ba^{2+}	1.3	0.17	7.41×10^3	1.2	23	51.3
Pb^{2+}	3.3	0.18	1.86×10^4	3.2	46	70.8
Hg^{2+}	4.0	15	2.63×10^2	1.6	33	47.9

[a] Liesegang *et al.*, 1977
[b] Liesegang *et al.*, 1976
[c] Rodriguez *et al.*, 1977
[d] Rodriguez *et al.*, 1978

mined by a conformational rearrangement of the ligand (and are therefore little dependent on the type of cation) leads to the interesting hypothesis that the selectivity of a crown ether is due to variations in $\Delta S^{\ddagger}_{32}$ (Liesegang *et al.*, 1978).

Rates of decomplexation (k_{32}) of cation complexes can also be determined by nmr spectroscopy on the cation. Rates of complex formation are then calculated from k_{32} and the binding constant. The results for several ligands, cations, and solvents are given in Table 20. Despite the wide variations, the rates of complex formation are all in the range 2×10^7 to 8×10^8 M^{-1} s^{-1}. In contrast, rates of decomplexation for crown-ether complexes span a much broader range: 6.1×10^2 to 2×10^5 s^{-1}. Comparison of crown-ether data with data for [2.2.2]-cryptand [37] and the linear polyether [92] also shows that the

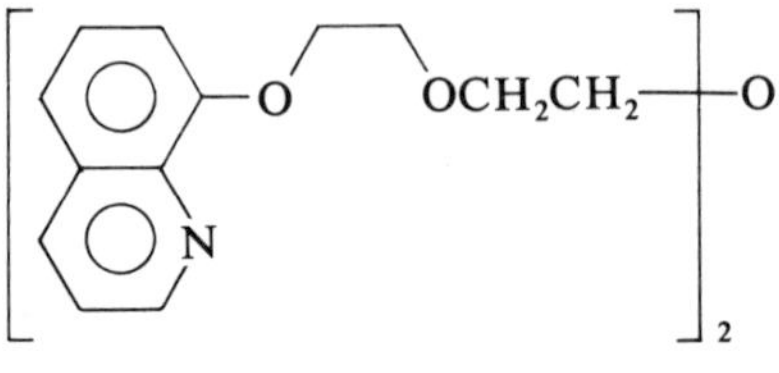

[92]

TABLE 20

Kinetic and thermodynamic parameters for metal-cation complexation

Ligand	Cation	Solvent	Temp. /°C	$k_{23}/M^{-1}\,s^{-1}$	k_{32}/s^{-1}	E_a /kcal mol^{-1}	K_1/M^{-1}	Ref.
18-Crown-6 [3]	Cs^+	Pyridine	25	—	9.3×10^3	8.3	$>10^6$	a
Dibenzo-18-crown-6 [11]	Na^+	DMF	25	$\sim 6 \times 10^7$	1.0×10^5	12.6	6×10^2	b
		MeOH	25	3.2×10^8	1.4×10^4	11.7	2.3×10^4	c
		Dimethoxyethane	25	6.5×10^7	1.5×10^4	13.3	4.5×10^3	c
cis-4,4′-Dinitro-[11]		DMF	25	2.3×10^7	2.0×10^5	12.5	1.15×10^2	c
cis-4,4′-Diamino-[11]		DMF	25	1.2×10^8	1.9×10^5	13.1	6.15×10^2	c
cis-4,4′-Dimethyl-[11]		THF	2	—	3.2×10^3	12.5	—	d
Dibenzo-18-crown-6 [11]	K^+	MeOH	−34	—	6.1×10^2	12.6	—	e
Dibenzo-30-crown-10 [14]	Na^+	MeOH	25	$>1.6 \times 10^7$	$>1.3 \times 10^5$	—	$(1.3 \pm 0.2) \times 10^2$	f
	K^+			$(6 \pm 2) \times 10^8$	$(1.6 \pm 0.5) \times 10^4$	—	$(3.7 \pm 0.4) \times 10^4$	f
	Rb^+			$(8 \pm 2) \times 10^8$	$(1.8 \pm 0.4) \times 10^4$	—	$(4.4 \pm 0.5) \times 10^4$	f
	Cs^+			$(8 \pm 1) \times 10^8$	$(4.7 \pm 0.6) \times 10^4$	—	$(1.7 \pm 0.2) \times 10^4$	f
	NH_4^+			$>3 \times 10^7$	$>1.1 \times 10^5$	—	$(2.7 \pm 0.3) \times 10^2$	f
	Tl^+			$(8 \pm 1) \times 10^8$	$(2.5 \pm 0.3) \times 10^4$	—	$(3.2 \pm 0.4) \times 10^4$	f
Dicyclohexyl-18-crown-6 [21]	Na^+	MeOH	25	2.6×10^8	5.2×10^4	8.3	4.8×10^3	c
[2.2.2]-Cryptand [37]	Na^+	MeOH	25	2.7×10^8	2.87	—	9.5×10^7	g
	K^+			4.7×10^8	1.8×10^{-2}	—	2.6×10^{10}	g
[92]	K^+	MeOH		$(1.1 \pm 0.3) \times 10^8$	$(4 \pm 1) \times 10^3$	—	3.2×10^4	h

[a] Mei *et al.*, 1977a
[b] Shchori *et al.*, 1971
[c] Shchori *et al.*, 1973
[d] Wong *et al.*, 1970
[e] Shporer and Luz, 1975
[f] Chock, 1972
[g] Cox *et al.*, 1978
[h] Tümmler *et al.*, 1977

rates of complex formation are very similar and that the differences in thermodynamic stability are almost fully accounted for by the differences in rates of decomplexation.

Shchori *et al.* (1971, 1973) have pointed to the remarkable similarity ($E_a = 12.6 \pm 0.7$ kcal mol^{-1}) of the activation energies for decomplexation of dibenzo-18-crown-6 [11] and substituted derivatives (Table 20). They too were of the opinion that the activation energy for decomplexation reflects the energy required for a conformational rearrangement. The fact that different crown ethers give different E_a values can also be explained by this model.

For a more extensive discussion of the kinetics of complexation with cryptands and natural antibiotics, the reader is referred to a recent review of Liesegang and Eyring (1978).

4 Chemical reactivity of metal cation complexes

GENERAL

As outlined in (1) at the beginning of Section 2, a solution containing a metal salt (MX) and a crown ether (Cr) may contain three different anionic species, namely, the free anion X^-, the tight ion pair $M^+.X^-$ and the crown ether-separated ion pair $M^+.Cr.X^-$. All of these species may react with a substrate SY with different rate constants. As the reaction proceeds, any leaving group Y may affect the rate of the reaction by competing with $M^+.X^-$ for crown ether as shown in (11). It is evident from (1) and (11) that the effect

$$M^+.Cr.X^- + M^+.Y^- \rightleftharpoons M^+.Cr.Y^- + M^+X^- \qquad (11)$$

of a crown ether on the rate of a reaction depends on several factors, such as its complexing ability towards M^+, $M^+.X^-$, $M^+.Y^-$, the degrees of dissociation of the ion pairs $M^+.X^-$, $M^+.Y^-$ and the corresponding crown ether-separated ion pairs, and the relative reactivities of free anion, ion pair and crown ether-separated ion pair.

It has become popular to attribute to crown ethers an "anion activation" effect which term suggests that their presence causes an increase in reaction rate. Although this is certainly true in many cases, it does not hold for reactions in which the ion pair is much more reactive than the free anion as a result of a cation-assisted reaction pathway. In addition to "anion activation", crown ethers also exhibit "cation deactivation".

Reactions performed under two-phase conditions are further complicated by the partitioning of the reactants and catalyst over the two phases. In the case of quaternary ammonium phase-transfer catalysis, the mechanistic aspects have received a great deal of attention (Brändström, 1977; Makosza, 1975; Starks and Owens, 1973). In contrast, the mechanism of crown ether-type phase-transfer catalysis has hardly been investigated at all, despite its well-established preparative value (Weber and Gokel, 1977; Starks and Liotta, 1978). Most of our present knowledge about the catalytic activity of crown ethers in two-phase systems therefore originates from preparative experiments.

NUCLEOPHILIC SUBSTITUTION

Homogeneous conditions

The characteristics of homogeneous crown-ether catalysis were nicely demonstrated by Thomassen *et al.* (1971) who studied the rate of alkylation of potassium phenoxide with 1-bromobutane in dioxan at 25°C. By measuring the initial consumption of phenoxide (τ_i in M s^{-1}), any effect of the bromide ion was neglected. The results (Table 21) show hardly any effect of tetraglyme

TABLE 21

Crown ether-catalysed reaction of 1-bromobutane (0.05 or 0.1 M) with potassium phenoxide (0.02 M) in dioxan at 25°C[a]

Catalyst (5.4%m)	$10^5\ \tau_i/[BuBr]_0$ s^{-1}	Catalyst (5.4%m)	$10^5\ \tau_i/[BuBr]_0$ s^{-1}
None	0.01		
$CH_3O(CH_2CH_2O)_4CH_3$ [43]	0.015	Dibenzo-18-crown-6 [11]	0.37
		4,4′-Di-t-butyl-[11]	0.33
Dibenzo-12-crown-4 [93]	0.023	Benzo-18-crown-6 [9]	1.00
Dibenzo-15-crown-5 [94]	0.023	Dicyclohexyl-18-crown-6 ([20] + [21])	2.80
Benzo-15-crown-5 [8]	0.043	Dibenzo-24-crown-8 [13]	0.075

[a] Thomassen *et al.*, 1971

[43] and crown ethers such as dibenzo-12-crown-4 [93] and dibenzo-15-crown-5 [94], having too small a cavity and binding sites of low basicity. The highest activity was found for dicyclohexyl-18-crown-6, which gave a 280-fold rate increase when present in 0.054 molar equivalents (based on phenoxide).

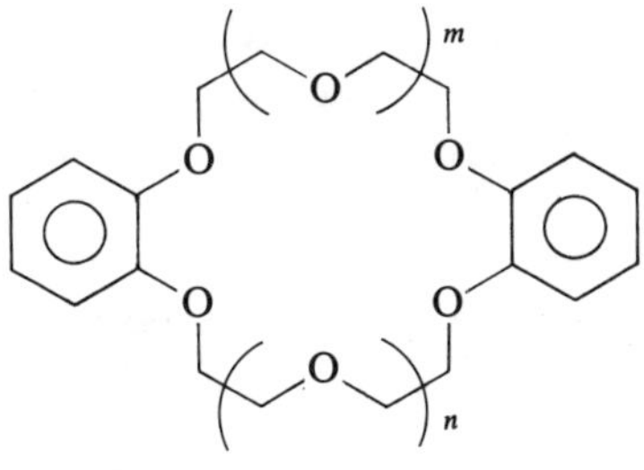

[93], $m = n = 0$

[94], $m = 0, n = 1$

The relative order of the catalytic activities of the crown ethers ([20] + [21]) > [9] > [11] > [13] > [8]) is the same as the relative order of their capacities to bind K^+ (Table 4). However, the intrinsic reactivities of the ion pairs were also dependent on crown-ether structure, as was shown by experiments in which the alkylation rates were determined at various crown/phenoxide molar ratios. The curve obtained (Fig. 2) is similar to the curves found in titration experiments (Live and Chan, 1976; De Jong *et al.*, 1976b), and shows that the rate constant reaches a maximum (called "plateau kinetics" in the literature) when all of the salt is complexed.

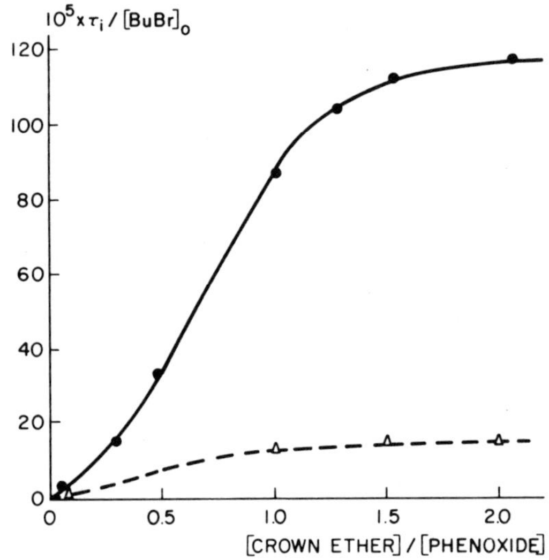

FIG 2 Effect of [crown ether]/[phenoxide] ratio on the alkylation of potassium phenoxide with 1-bromobutane at 25°C. ——, dicyclohexyl-18-crown-6; ---, di(t-butylbenzo)-18-crown-6. Data taken from Thomassen *et al.* (1971)

The behaviour of the dibenzo-18-crown-6 derivative is similar, yet the highest attainable rate is only one eighth of that observed with dicyclohexyl-18-crown-6, which points to an important difference in reactivity between crown ether-separated ion pairs. Compared to tetrabutylammonium phenoxide, the dicyclohexyl-18-crown-6/K^+ phenoxide was 2.6 times less reactive. The addition of 0.05 M dicyclohexyl-18-crown-6 to dioxan resulted in the alkylation rate constant becoming the same as that observed in pure tetraglyme.

Hirao *et al.* (1978a,b) have reported similar results. The rate of reaction of sodium phenoxide with 1-bromobutane was 88 times faster in dimethoxyethane than in dioxan. The presence of 0.135 M dibenzo-18-crown-6 [11] in dioxan completely levelled out the difference with dimethoxyethane. It is interesting to note that these authors found polyethylene oxide (PEO) of molecular weight 3.3×10^4 to be just as effective as dibenzo-18-crown-6. The activity of PEO was greatly affected by the molecular weight, being very low for glymes, and increasing steadily up to a molecular weight of *ca.* 10^4 where a plateau was reached. Other polyethers, such as poly(oxetane) or poly-(tetrahydrofuran) were ineffective as catalysts, probably because the binding sites are too far apart.

The arylation of potassium phenoxide has been studied by Litvak and Shein (1976). In the absence of crown ether the rate of reaction with *p*-nitrobromobenzene in various solvents increases in the order: dioxan < methanol < pyridine < DMSO. In the presence of either 18-crown-6 [3] or dibenzo-18-crown-6 [11], the order of the reaction rates in dioxan and methanol is reversed (Table 22). The effect of crown ether addition on the rate

TABLE 22

Reaction of potassium phenoxide with *p*-nitrobromobenzene in the presence of 18-crown-6 [3] and dibenzo-18-crown-6 [11] at 100°C[a]

	$10^3 k_2/M^{-1}\,s^{-1}$		
Solvent	No additive	[3]	[11]
Dioxan[b]	0.014	3.7	2.4
Methanol[c]	0.28	0.31	0.37
Pyridine[c]	6.8	21	57
DMSO[c]	110	82	110

[a] Litvak and Shein, 1976
[b] With two-fold excess of crown ether (based on KOPh); plateau values
[c] With equimolar amount of crown ether (based on KOPh)

of the reaction increases with decreasing cation-solvating power of the solvent: methanol ≈ DMSO < pyridine < dioxan. Hardly any effect is found in good cation-solvating solvents such as methanol and DMSO. The plateau values for [3] and [11] in dioxan are again different, 18-crown-6 forming the more reactive complex.

The small effect on reaction rates of the addition of crown ethers to the lower alcohols was also observed in the reaction of potassium acetate with 1-bromobutane in ethanol (Hirao *et al.*, 1978a,b). The displacement of fluorine in either *o*-nitro- or *p*-nitro-fluorobenzene by a methoxy group, by reaction with potassium methoxide in methanol was hardly influenced by the presence of dicyclohexyl-18-crown-6 (Del Cima *et al.*, 1973). Mariani *et al.* (1978), too, found no effect of dicyclohexyl-18-crown-6 on the rate of displacement of chlorine in 1-chloro-2,4-dinitrobenzene by reaction with potassium methoxide in methanol. The effect of crown ether becomes noticeable in benzene–methanol mixtures containing at least 50% (by volume) of the former constituent. In 99% benzene–1% methanol a 300-fold increase in rate was observed upon addition of an equimolar amount of the crown ether.

In sharp contrast to the $KOCH_3/CH_3OH$ system, the KOt-Bu/t-BuOH combination is quite sensitive to the presence of cation-complexing agents, as was shown by Del Cima *et al.* (1973). Whereas on addition of ([20] + [21]), the rate of displacement of fluorine by reaction with KOt-Bu increased by only a factor of 3 in the case of 1-fluoro-2-nitrobenzene, it increased by a factor of 1.4×10^3 for 1-fluoro-4-nitrobenzene (Table 23). Consequently, the

TABLE 23

Reaction of fluoronitrobenzenes with potassium alkoxide (KOR) in ROH in the presence of dicyclohexyl-18-crown-6 ([20] + [21]) at 31°C[a]

Substrate	Nucleophile	([20] + [21]) /%m	$10^4 k_2$ /M^{-1} s^{-1}	k_O/k_P
1-Fluoro-2-nitrobenzene	$KOCH_3$	—	5.0	1.0
		104	4.2	0.6
	KOt-Bu	—	54	360
		100	170	0.8
	$PhCH_2N(CH_3)_3$Ot-Bu	—	170	7.4
1-Fluoro-4-nitrobenzene	$KOCH_3$	—	4.7	—
		104	7.0	—
	KOt-Bu	—	0.15	—
		100	220	—
	$PhCH_2N(CH_3)_3$Ot-Bu	—	23	—

[a] Del Cima *et al.*, 1973

ortho/para reactivity ratio changes from 360 in the absence, to 0.8 in the presence of crown ether. The authors rationalized these results by assuming that the *o*-nitro group can stabilize the intermediate Meisenheimer complex [95], whereas the *p*-nitro group cannot. Addition of a strong cation-binding agent removes the cause of the stabilization of the transition state, so the *ortho/para* reactivity ratio approaches unity. In the case of $KOCH_3$, the free methoxide anion is thought to be the reactive species and this cannot give [95].

The reactivity of complexed KOt-Bu (i.e. complexed by a mixture of [20] and [21]) towards 1-fluoro-2-nitrobenzene is almost the same as that of the quaternary ammonium salt (Table 23). Sam and Simmons (1974) also observed comparable reactivities of KBr or KI complexed by [20] and the

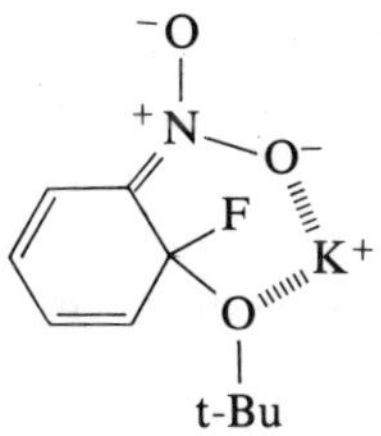

[95]

corresponding tetrabutylammonium salts in their reactions with n-butyl brosylate in acetone at 25 °C. The second-order rate constants for the two isomeric dicyclohexyl-18-crown-6 complexes with KI were only slightly different, 2.08×10^{-3} and 2.32×10^{-3} M^{-1} s^{-1} for the *cis-syn-cis* [20] and the *cis-anti-cis* [21] isomer, respectively.

The effect of crown ethers on the alkylation of sodium diethyl n-butylmalonate by 1-bromobutane has been studied by Zaug *et al.* (1972). The absence of a common-ion rate depression in dimethylformamide (DMF) pointed to an ion pair being the kinetically active species. The addition of dicyclohexyl-18-crown-6 (a mixture of [20] and [21]) accelerates the alkylation in both benzene and tetrahydrofuran (THF) (Table 24). The rates reach a plateau, indicating that at a crown-ether concentration of 0.5 M the ion pair is fully converted to the crown ether-separated ion pair which is slightly less reactive than the uncomplexed ion pair in DMF. The rate constant in pure dimethoxyethane (DME) is equal to that observed in THF or benzene

TABLE 24

Second-order rate constants for the reaction of sodium diethyl n-butylmalonate with 1-bromobutane at 25 °C[a]

Solvent	Additive (M)	$10^5 k_2$ /M^{-1} s^{-1}	Solvent	Additive (M)	$10^5 k_2$ /M^{-1} s^{-1}
DMF	—	323	THF	—	5
				[20] + [21], 0.01	13.5
DME	—	29		0.02	19.7
				0.04	29.7
Cyclohexane	DMF, 0.308	2.9		0.10	55.3
	0.681	7.6		0.50	84.7
	1.194	15.9			
	3.293	53.1	Benzene	[20] + [21], 0.0116	5.72
	DME, 0.301	0.93		0.0228	17.0
	0.684	1.48		0.0406	32.7
	0.954	1.84		0.1015	63.6
	2.820	9.11		0.527	138.1

[a] Zaug *et al.*, 1972

containing as little as 0.04 M crown ether. The excellent cation-complexing ability is also evident from a comparison with the alkylation rate in cyclohexane to which DMF or DME has been added as cation-solvating species (Table 24): as compared with crown ethers, these additives not only give a far less spectacular rate enhancement, but also fail to produce plateau values for k_2.

The effect of crown ethers on the rates and stereochemistry of the alkylation of metal acetoacetates has been studied by Cambillau *et al.* (1976, 1978) and Kurts *et al.* (1973, 1974). Since the enolate can adopt various conformations ([96]–[99]), O-alkylation may produce either the *cis* ([100]) or the *trans* ([101]) isomer, whereas C-alkylation affords [102]. The reaction of the sodium

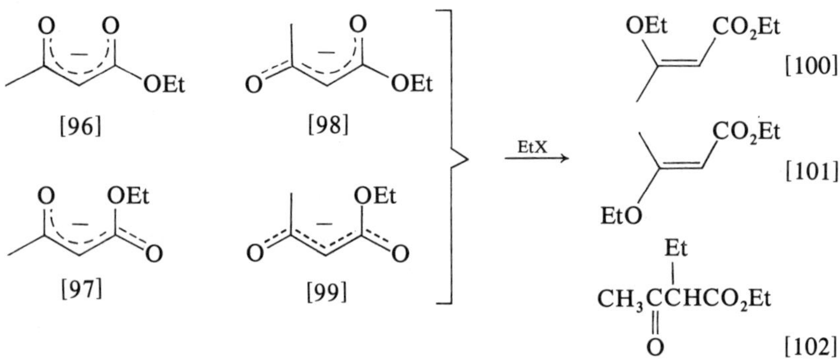

(12)

derivative with ethyl tosylate in tetrahydrofuran at 50°C gives predominantly C-alkylation (Table 25). In the presence of 18-crown-6 [3], the rate of O-alkylation increases by a factor of 80 but that of C-alkylation only by a factor of 8. An even more pronounced effect is found with [2.2.2]-cryptand [37], for which these factors are 1.9×10^3 and 39 respectively. Hence, the use of cation-complexing agents shifts the product distribution from C- to O-alkylates. The magnitude of this effect depends on the type of leaving group, cation, solvent and ligand (Table 25).

The proportion of C-alkylation increases in the order OTs $<$ Br $<$ I, a sequence which is often associated with the balance of hardness between nucleophile and nucleofuge (Smith and Hanson, 1971). The work of Kurts *et al.* (1974) indicates that the overall reaction rate of the crown ether-assisted alkylation increases in the order $Na^+ < K^+ < Rb^+ < Cs^+$, which, according to these authors, reflects the increasing distance between cation and anion in the ion pairs. The high reactivity of the tetraphenylarsenate also fits in with this picture. The decrease of the k_C/k_O ratio is only small in good cation-solvating solvents such as dimethyl sulfoxide (DMSO). Alkylation of the sodium derivative of [103] with ethyl iodide in DMSO gave $k_C/k_O = 15.7$; addition of

TABLE 25

Crown ether-assisted alkylation of metal ethyl acetoacetates

Cation	Substrate	Solvent	Temp. /°C	Crown ether	$10^3k_2{}^d/M^{-1}\,s^{-1}$ Total	C	O	k_C/k_O	% O-*trans*	Ref.
Na^+	EtOTs	THF	50	—	0.12	0.11	0.012	9.0	<10	[a]
				[3]	1.75	0.87	0.87	1.0	85	[a]
				[37]	27.5	4.3	23.2	0.19	100	[a]
K^+	EtOTs	THF	25	[3]	1.2	0.38	0.78	0.49	79	[b]
				[37]	32.5	3.7	28.2	0.13	100	[b]
	EtBr	THF	25	[3]	2.7	2.4	0.4	7.25	90	[b]
				[37]	131.7	75.6	56.0	1.35	100	[b]
	EtI	THF	25	—	0.6	—	—	—	—	
				[3]	66.7	64.8	1.9	33.5	86	[b]
				[37]	2500	1980	520	3.8	100	[b]
Na^+	EtOTs	Dioxan	25	([20] + [21])	0.063	0.044	0.019	2.3	62	[c]
K^+	EtOTs	Dioxan	25	—	$<10^{-4}$	—	—	—	—	
				([20] + [21])	0.27	0.11	0.16	0.7	70	[c]
Rb^+	EtOTs	Dioxan	25	([20] + [21])	0.44	0.23	0.21	1.1	70	[c]
Cs^+	EtOTs	Dioxan	25	([20] + [21])	0.93	0.46	0.47	1.0	89	[c]
$(C_6H_5)_4As^+$	EtOTs	Dioxan	25	—	7.2	1.5	5.,7	0.26	100	[c]

[a] Cambillau *et al.*, 1976
[b] Cambillau *et al.*, 1978
[c] Kurts *et al.*, 1974
[d] Plateau values of second-order rate constants

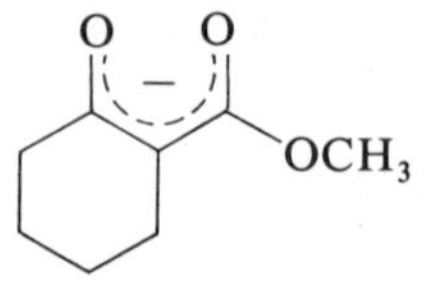

[103]

dicyclohexyl-18-crown-6 and [2.2.2]-cryptand decreased k_C/k_O to 13.3 and 9 respectively (Neé and Tchoubar, 1976). Alkylation with ethyl tosylate was hardly affected by the presence of cation-complexing agents. The authors concluded that the free anion is the dominant species in DMSO solution. Akabori *et al.* (1978) also found that crown ethers have no effect on the alkylation of sodium phenoxide in ethanol solution, whereas in apolar solvents such as diethyl ether or benzene the relative extent of O-alkylation increases upon addition of crown ethers.

The proportion of the *trans*-O-alkylated product [101] increases in the order: no ligand $\ll$ 18-crown-6 < [2.2.2]-cryptand. This difference was attributed to the fact that the enolate anion in a crown-ether complex is still capable of interacting with the cation, which stabilizes conformation [96]. For the cryptate, however, cation–anion interactions are less likely and electrostatic repulsion will force the anion to adopt conformation [99], which is the same as that of the free anion in DMSO. This explanation was substantiated by the fact that the anion was found to have structure [96] in the solid state of the potassium acetoacetate complex of 18-crown-6 (Cambillau *et al.*, 1978). Using ^{23}Na NMR, Cornelis *et al.* (1978) have recently concluded that the active nucleophilic species is the ion pair formed between 18-crown-6 and sodium ethyl acetoacetate, in which Na^+ is co-ordinated to both the anion and the ligand.

The preferential stabilization of conformation [96] by the cation is annulled upon binding of the cation with a cryptand. Similar effects have been observed with crown ethers. The sodium salt of acac in methanol at −58°C consists of 35% [104] and 65% [105]. The addition of 18-crown-6 [3] shifts the

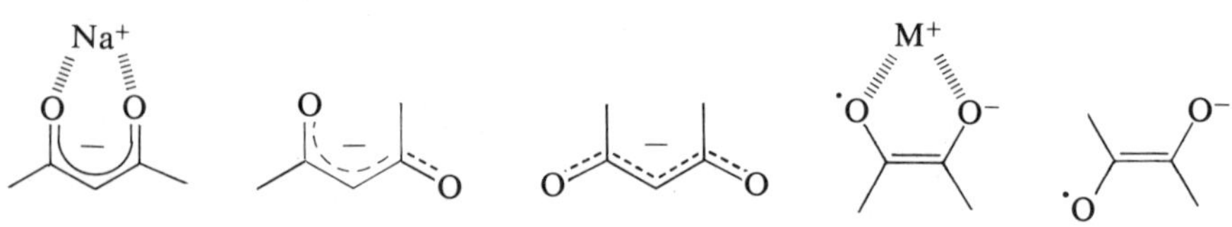

[104] [105] [106] [107] [108]

equilibrium completely towards the free anion [105] (Raban *et al.*, 1977). In pyridine solution, the equilibrium was shifted towards [106] (Noe and Raban, 1974). Such effects were also noted for the conformational equilibrium of semidione radical anions (Russell *et al.*, 1978). The *trans*/*cis* ratios

[108]/[107] in DMSO at 25°C were 1.2 and 13 for the sodium and potassium dimethylsemidiones, respectively. The addition of a slight molar excess of [2.2.2]-cryptand [37] converted the *cis*-chelated radical anion [107] into the free *trans* anion [108]. With the aid of either 15-crown-5 or 18-crown-6, *trans* was also converted into *cis*, but even an excess of ligand failed to bring about complete conversion. It was concluded that the crown ether-complexed cation can still chelate the *cis* radical anion [107]. In t-butyl alcohol solution, the conversion of *trans* into *cis* is also complete for 15-crown-5 and 18-crown-6.

The second-order rate constant for the methylation of sodium 9-fluorenone oximate in 33.5% acetonitrile/66.5% t-butyl alcohol solution was found to decrease with increasing concentration of the salt, suggesting an equilibrium (13) between the reactive free anion [109] and the less reactive ion pair [110]

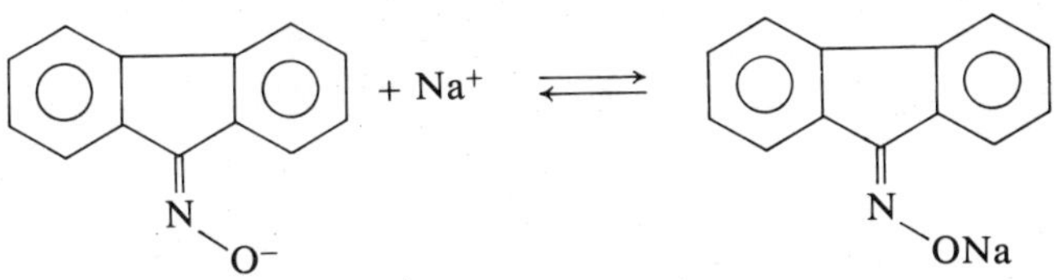

(13)

[109] [110]

(Smith and Hanson, 1971). Upon addition of $NaBPh_4$, the free anion was converted into the ion pair, and upon addition of dibenzo-18-crown-6 [11] the ion pair was converted into the free anion. The results (Table 26) show the free anion to be much more reactive than the ion pair towards both methyl iodide and methyl tosylate. The increase in rate is again mainly due to the enhanced reactivity of the hard oxygen centre. The ratio of the rates of alkylation at the two sites of the ambident anion (k_N/k_O) decreased from about 1.2 for the ion pair to 0.55 and almost zero for the free anion. Increased alkylation of the hard

TABLE 26

Methylation of sodium fluorenone oximate in 33.5% acetonitrile and 66.5% t-butyl alcohol at 25°C in the presence of $NaBPh_4$ or dibenzo-18-crown-6 [11][a]

Substrate	Additive	$10^2k_2/M^{-1}\,s^{-1}$ Total	N	O	k_N/k_O
CH_3I	$NaBPh_4$	1.4	0.8	0.6	1.3
	[11]	99	34	65	0.55
CH_3OTs	$NaBPh_4$	1.5	0.8	0.7	1.2
	[11]	68	<1	68	<0.01

[a] Smith and Hanson, 1971

oxygen centre of sodium 2-naphthoxide by reaction with benzyl bromide in THF and benzene also occurred upon addition of benzo-18-crown-6 [9] or [2.2.2]-cryptand [37], the latter being the more effective (Akabori and Tuji, 1978).

The difference in behaviour between crown ethers and cryptands is also evident from the results reported for the reaction of $AgNO_3$ with acetobromoglucose [111] (14). In the presence of equimolar amounts of di-

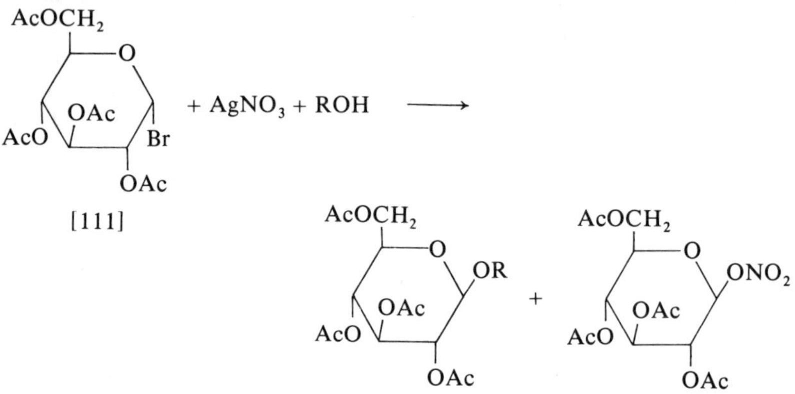

[111] [112] [113] (14)

benzo-18-crown-6 [11] (relative to Ag^+), the only product is the β-glucoside [112] (Knöchel *et al.*, 1974). The reactivities of the alcohols decrease in the order methanol > isopropyl alcohol > t-butyl alcohol > cyclohexanol. The use of [2.2]-cryptand [86] in t-butyl alcohol solution gave 95.1% [112] and 4.9% [113], whereas [2.2.2]-cryptand [37] produced 63.6% [112] and 36.6% [113] (Knöchel and Rudolph, 1974). The authors conclude that in contrast to [2.2.2]-cryptand the crown ethers do not "activate" the nitrate anion to such an extent that it can successfully compete with the alcohol as a nucleophile.

The acceleration of the reaction between benzyl chloride and metal acetates in acetonitrile has been studied using several crown ethers (Dorn *et al.*, 1977; Knöchel *et al.*, 1975). Very surprisingly, the overall reaction was found to be first order in the presence of cryptands and second order in the presence of crown ethers. The order in benzyl chloride was one in both cases, indicating a zero-order dependence on acetate concentration for the cryptand-catalysed reaction. Unfortunately, the authors did not give all the details of the experiments on which these conclusions were based. Hence, no explanation can be offered for the apparent discrepancy between their results and those of Cambillau *et al.* (1976, 1978), who observed a first-order dependence on the concentration of metal acetoacetates in alkylations, irrespective of whether or not they were performed in the presence of crown ethers or cryptands.

The data obtained with various acetates and ligands are collected in Table 27. For comparison, Dorn *et al.* (1977) have converted the *pseudo*-first-order rate constants for cryptates into second-order rate constants. For all the cations studied, but particularly for Na^+ and Ag^+, cryptands [36] and [37] are more effective than crown ethers. In the case of Ag^+, this is undoubtedly due to the stability of its cryptates being higher than that of its crown-ether complexes. In the case Na^+, the gap between macrocyclic and macrobicyclic ligands is surprisingly wide, particularly considering the fact that association constants of Na^+/crown-ether complexes in acetonitrile are high ($\sim 10^5$ M^{-1}) and comparable to K^+ binding constants (Table 17). Furthermore, Dorn *et al.* (1977) find 15-crown-5 to be the most efficient crown ether for the reaction between potassium acetate and benzyl chloride, whereas Thomassen *et al.* (1971) find it to be far less efficient than 18-crown-6 derivatives in the alkylation of potassium phenoxide (Table 21). Further studies are needed to show the relative effects of binding constants, dissociation constants and complex reactivities.

Liotta and Grisdale (1975) have reported on the relative nucleophilicities of anions whose potassium salts were solubilized into acetonitrile by 18-crown-6 [3]. The results (Table 28) show the sequence $N_3^- > OAc^- > CN^- > F^- > Cl^- \approx Br^- > I^- > SCN^-$, which is very different from the reactivity scale in water: $CN^- > I^- > SCN^- > N_3^- > Br^- > Cl^- > OAc^- > F^-$. Furthermore, the relative nucleophilicities in acetonitrile vary only by a factor of 30, whereas in water they differ by as much as a factor of 1000. The fact that gas-phase nucleophilicities also span a narrow range led the authors to conclude that anion solvation is much less important in acetonitrile than in water. The values recently reported by Lemmetyinen *et al.* (1978) for the relative nucleophilicities of anions towards methyl methanesulfonate in benzene show the same sequence as in protic solvents, however. The authors offered no explanation for this peculiar behaviour.

It is interesting to compare the nucleophilicity sequence found by Liotta and Grisdale (1975) for crown ethers, with the sequence recently reported by Landini *et al.* (1978) for 'onium compounds. For hexadecyltributylphosphonium as the cation, the reactivity sequence (for n-octyl methanesulfonate) in chlorobenzene was $CN^- > N_3^- > Cl^- > Br^- > I^- > SCN^-$ (Table 28), which is the same as that observed in DMSO. In protic solvents such as methanol, an entirely different sequence is observed, and the reactivity range in methanol is *smaller* than in aprotic solvents. The same picture emerges from a comparison of the data obtained in anhydrous chlorobenzene with those obtained under liquid–liquid phase-transfer conditions. In the latter case, Landini *et al.* found the anions to be partially hydrated, resulting in the same nucleophilicity sequence as in protic solvents but with far less variation. Except for a reversal of N_3^- and CN^-, the behaviour of the crown ether-complexed potassium ion is

TABLE 27

Second-order rate constants ($10^3 k_2/M^{-1}\ s^{-1}$) for the homogeneous reaction of benzyl chloride with metal acetates in acetonitrile at room temperature[a]

Ligand[b]	Cation				
	Na^+	K^+	Cs^+	Ag^+	NH_4^+
[2.2.2]-Cryptand [37]	4.37	5.58	4.49	1.23	1.52
[2.2.1]-Cryptand [36]	3.90	2.27	2.46	4.53	0.38
Dicyclohexyl-18-crown-6 ([20] + [21])	6.7×10^{-4}	0.75	0.31	5.93×10^{-3}	[c]
18-Crown-6 [3]	2.67×10^{-4}	0.53	0.80	[c]	[c]
Dibenzo-18-crown-6 [11]	—	0.074	—	[c]	[c]
15-Crown-5 [2]	5.9×10^{-3}	0.79	—	[c]	[c]
Benzo-15-crown-5 [8]	3.1×10^{-3}	0.29	—	[c]	[c]
Dibenzo-24-crown-8 [13]	—	0.01	0.27	[c]	[c]
None	1.0×10^{-5}	1.6×10^{-5}	8.4×10^{-5}	1.2×10^{-5}	[c]

[a] Dorn *et al.*, 1977; Knöchel *et al.*, 1975

[b] Present in equimolar amounts

[c] No reaction

TABLE 28

Relative nucleophilicities of anions under homogeneous (H) and liquid–liquid phase-transfer (L,L-PT) conditions

Substrate	Solvent	Conditions	Temp. /°C	Cation	Relative rate constants								Ref.
					OAc^-	N_3^-	CN^-	F^-	Cl^-	Br^-	I^-	SCN^-	
n-$C_8H_{17}OSO_2CH_3$	Chlorobenzene	H	60	$C_{16}H_{33}P^+Bu_3$	—	23.5	29.0	—	6.6	2.7	1.0	0.3	a
n-$C_8H_{17}OSO_2CH_3$	Chlorobenzene	L,L-PT	60	$C_{16}H_{33}P^+Bu_3$	—	6.8	4.2	—	0.6	1.1	1.0	0.2	a
n-$C_8H_{17}OSO_2CH_3$	DMSO	H	60	$C_{16}H_{33}P^+Bu_3$	—	25.5	63.8	—	6.8	4.3	1.0	0.5	a
n-$C_8H_{17}OSO_2CH_3$	MeOH	H	60	$C_{16}H_{33}P^+Bu_3$	—	1.03	0.69	—	0.15	0.38	1.0	0.29	a
$C_6H_5CH_2OTs$	Acetonitrile	H	30	K^+/18-crown-6	9.6	10.0	2.4	1.4	1.3	1.3	1.0	0.3	b
$CH_3SO_2OCH_3$	Benzene	H	25	K^+/18-crown-6	0.14	—	1.4	0.005	0.015	—	—	—	c
				K^+/dicyclohexyl-18-crown-6	—	—	—	—	—	0.46	1.0	0.036	c

[a] Landini *et al.*, 1978
[b] Liotta and Grisdale, 1975
[c] Lemmetyinen *et al.*, 1978

very similar to that of 'onium compounds, which underlines the conclusion of Liotta and Grisdale (1975) that crown ether-solubilized anions are only weakly solvated.

The effect of solvation on the α-effect has been studied by Curci and Di Furia (1975) for the reactions of t-BuOK and t-BuOOK with *p*-nitrophenyl diphenylphosphinate [114] and *p*-nitrophenyl benzoate [115]. In water the α-effects (rate ratios for t-BuOO$^-$ and $CF_3CH_2O^-$) were 6.6 for [114] and 5.5

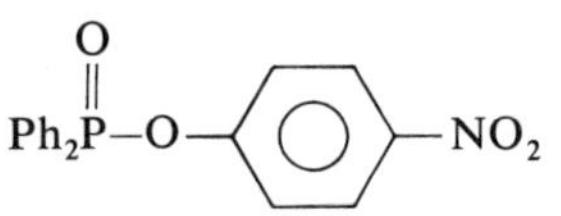

[114]

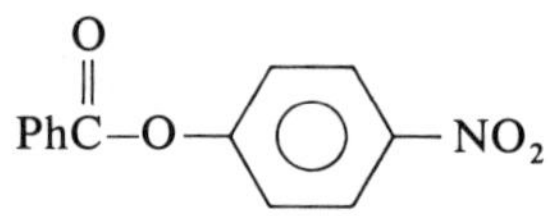

[115]

for [115]. In toluene, in the presence of dicyclohexyl-18-crown-6 (mixture of [20] and [21]), the α-effects (rate ratios for t-BuOO$^-$ and t-BuO$^-$) were 2.7 and 5.2 for [114] and [115] respectively. The observation that the effect does not disappear in toluene was presented as evidence that solvation is not a major factor determining the α-effect.

Two-phase liquid–solid reactions

Knöchel *et al.* (1977a) have studied the effect of the structure of the ligand on its ability to catalyse the reaction between solid metal acetates and benzyl chloride dissolved in acetonitrile. Approximate half-lives for the reactions are given in Table 29. For crown ethers, the reactivity sequence decreases in the

TABLE 29

Times (h) needed to reach 50% conversion in the reaction of benzyl chloride with metal acetates in acetonitrile at *ca.* 25°C under solid–liquid phase-transfer conditions[c]

Catalyst[a]	Na^+	K^+	Cs^+	Ag^+	NH_4^+
[2.2.2]-Cryptand [37]	25	1.3	23	2	19
[2.2.1]-Cryptand [36]	16	2.5	3.5	2.2	34
[2.2]-Cryptand [86]	200	73	110	19	—
Dicyclohexyl-18-crown-6 ([20] + [21])	92	2	16	[b]	63
18-Crown-6 [3]	165	3	15	[b]	38
Dibenzo-18-crown-6 [11]	—	6	47	[b]	—
15-Crown-5 [2]	—	6	24	[b]	—
Benzo-15-crown-5 [8]	—	5	35	[b]	—
Dibenzo-24-crown-8 [13]	—	>200	—	—	—

[a] 0.1 Molar equivalent
[b] No reaction
[c] Knöchel *et al.*, 1977a

order $K^+ > Cs^+ > NH_4^+ > Na^+ \gg Ag^+$. The excellent ability of cryptands to complex Ag^+ is apparent from the reactivity sequence $K^+ > Ag^+ > Cs^+, Na^+, NH_4^+$ for [2.2.2]-cryptand [37]. For potassium acetate, the sequence of crown ether activities ([20] + [21]) > [3] > [8] > [2], [11] is quite different from the sequence for the same reaction under homogeneous conditions (Table 27). The authors suggest that mass transport has become a limiting factor in this particular solid–liquid system. Clearly, the dissolution process requires a closer study, particularly the question of whether or not anion exchange is accompanied by cation exchange.

It is worth mentioning at this point that according to Normant *et al.* (1975) simple polyamines such as tetramethylethylenediamine (TMEDA) are even more active than [2.2.2]-cryptand in the benzylation of acetates in acetonitrile under liquid–solid conditions. These authors suggested that the activity was due to salt solubilization by cation complexation and not to formation of a quaternary ammonium ion since the latter showed no activity. This statement, however, is not in line with the results of Cote and Bauer (1977), who were unable to detect any interaction between K^+ and TMEDA in acetonitrile. Furthermore, Vander Zwan and Hartner (1978) found Aliquat 336 (tricaprylylmethylammonium chloride) to be almost as effective as TMEDA in this reaction (Table 30). It might well be, however, that in amine-catalysed benzylation reactions the quaternary salt formed *in situ* acts both as a reactant and as a phase-transfer catalyst, since Dou *et al.* (1977) have shown that the benzyltriethylammonium ion is a powerful benzylation agent.

The data in Table 30 show an interesting difference between quaternary ammonium and amine on the one hand and crown ether on the other; whereas for F^- and AcO^- the differences are small, for CN^- the crown ether is a far more efficient catalyst. The authors suggested that the observed difference

TABLE 30

Reaction of potassium salts with alkyl halides in acetonitrile under solid–liquid conditions[b]

Substrate	Anion	Temp./°C	$t_{1/2}$[a]/h Aliquat 336	18-crown-6	TMEDA
$PhCH_2Cl$	F^-	83	42	147	107
	CN^-	83	2.2	0.03	1.1
	AcO^-	25	0.75	1.15	1.3
$1\text{-}C_6H_{13}Br$	CN^-	83	0.98	0.57	1.26
		25	34	0.33	27
	AcO^-	83	0.11	0.13	0.15

[a] Time required for 50% conversion; 0.1 molar equivalent of catalyst
[b] Vander Zwan and Hartner, 1978

might be due to the unfavourable apparent extraction constant of CN^- by quaternary salts. Lehmkuhl *et al.* (1977) found that linear polyethers such as octaglyme are only slightly less active than 18-crown-6 in the reaction of benzyl chloride and potassium fluoride, acetate, and cyanide under solid–liquid conditions.

The relative reactivities of anions and of leaving groups have been studied by Liotta and co-workers (Table 31). The anion reactivity follows the same order

TABLE 31

Reaction of potassium salts with alkyl halides in acetonitrile catalysed by 18-crown-6 [3] under liquid–solid conditions

Substrate	(conc./M)	Anion	Catalyst (conc./M)	Temp. /°C	$t_{1/2}$[a] /h	Ref.
$PhCH_2Br$	(2.0)	F^-	0.19	83	11.5	b
	(1.8)	CN^-	0.146	25	13	c
	(3.4)	AcO^-	0.16	25	2	d
$PhCH_2Cl$	(1.8)	CN^-	0.47	25	0.08	c
1-$C_8H_{17}Br$	(1.16)	F^-	0.19	83	115	b
	(1.4)	OAc^-	0.10	83	3	d
1-$C_6H_{13}Br$	(1.4)	OAc^-	0.10	25	150	d
	(1.8)	CN^-	0.138	83	10	c
1-$C_6H_{13}Cl$	(1.8)	CN^-	0.139	83	0.5	c
2-$C_8H_{17}Br$	(2.0)	CN^-	0.119	83	14	c
2-$C_8H_{17}Cl$	(2.0)	CN^-	0.122	83	59	c

[a] Time required for 50% conversion
[b] Liotta and Harris, 1974
[c] Cook *et al.*, 1974
[d] Liotta *et al.*, 1974

as found under homogeneous conditions: $AcO^- > CN^- \gg F^-$. Leaving group reactivities towards acetate and fluoride decrease in the order: $Br^- > OTs^- > Cl^-$. With cyanide, secondary bromides also react faster than chlorides, but interestingly enough, primary chlorides react much faster than the corresponding bromides. The authors think that this effect might be due to the difference between the KCl- and KBr-solubilizing capacities of a catalytic amount of crown ether. The high reactivity of primary chlorides compared to that of bromides is not restricted to reactions performed in acetonitrile; Shaw *et al.* (1978) recently reported that potassium cyanide reacted faster with 1-bromohexane than with 1-chlorohexane in hexamethylphosphoramide (HMPA) solution. On addition of 18-crown-6, the order of their reactivities was reversed.

The effect of solvent on crown ether-catalysed solid–liquid reactions has not been studied systematically. Observations made by Liotta and Harris (1974), Padwa and Dehm (1975), and Dupon Durst (1974) suggest that the reactions proceed more rapidly in acetonitrile than in benzene. In preparative experiments both solvents have been used to solubilize sodium azide (Nakajima *et al.*, 1977), potassium carboxylates (Dupont Durst, 1974; Akabori and Ohtomi, 1975; Padwa and Dehm, 1975), potassium phenoxide (Rall *et al.*, 1976) and potassium fluoride (Akhmetova *et al.*, 1978; Markezich *et al.*, 1977) for nucleophilic substitution reactions. Potassium nitrite reacts with alkyl halides under crown ether-catalysed solid–liquid conditions to form the corresponding nitro compounds (Zubrick *et al.*, 1975). The solvent for this reaction was not mentioned.

Two-phase liquid–liquid reactions

Landini *et al.* (1974, 1975) have studied nucleophilic substitutions in n-octyl derivatives in two-phase systems (substrate–water) catalysed by a mixture of dicyclohexyl-18-crown-6 isomers ([20] + [21]). They observed the following order of nucleophilicity: $I^- > CN^-, Br^- > Cl^- > F^-$ (Table 32), which is

TABLE 32

Nucleophilic substitutions in n-octyl derivatives catalysed by dicyclohexyl-18-crown-6 ([20] + [21])[a] under liquid–liquid phase-transfer conditions[d]

Substrate	Reagent[b]	Temp. /°C	Time /h	Yield /%
n-$C_8H_{17}Br$	KF	150	48	0
	KCl	80	45	60
	KI	80	3	100
	KSCN	80	1	100
	K_2S	80	12	75
	KSPh	40	2	100
n-$C_8H_{17}I$	KBr	80	3	40[c]
n-$C_8H_{17}OSO_2CH_3$	KF	115	42	65
	KCl	100	3	89
	KBr	100	0.5	96
	KI	100	0.11	100
	KCN	100	0.3	90
	NaCl	100	3	75
	NaBr	100	0.5	88
	NaI	100	0.16	100

[a] 0.05 Molar equivalent
[b] Saturated solution
[c] Same value after 26 h
[d] Landini *et al.*, 1974, 1975

reminiscent of the behaviour of the anions in water. In contrast to catalysis by phosphonium salts (Table 28), the cyanide is less reactive than iodide under these conditions. The reaction between n-octyl iodide and potassium bromide does not proceed to completion. This phenomenon is not uncommon in quaternary ammonium-catalysed reactions of alkyl iodides, and can be attributed to catalyst poisoning by the iodide anion produced (Dehmlow, 1977). It is not clear yet whether the effect in the crown ether-catalysed reaction has the same origin. An interesting difference between 18-crown-6 and dicyclohexyl-18-crown-6 shows up upon comparing the data of Tables 32 and 33; whereas the reaction between n-octyl bromide and potassium iodide

TABLE 33

Effect of catalyst structure on nucleophilic substitutions of n-$C_8H_{17}Br$ under liquid–liquid phase-transfer conditions[d]

Reagent[a]	Crown ether[b]	Temp. /°C	Time /h	Yield /%
KI	—	80	24	<4
KI	18-crown-6 [3]	80	3	35[c]
KI	Dibenzo-18-crown-6 [11]	80	40	80
KI	Dicyclohexyl-18-crown-6 ([20] + [21])	80	3	100
NaI	Benzo-15-crown-5 [8]	80	21	80
NaCl	[8]	100	24	0
NaCN	[8]	100	24	0

[a] Saturated aqueous solution
[b] 0.05 Molar equivalent
[c] Same value after 6 h
[d] Landini *et al.*, 1975

proceeds to completion when catalysed by dicyclohexyl-18-crown-6, it gives only 35% product when catalysed by 18-crown-6. From this observation the authors (Cinquini and Tundo, 1976) conclude that 18-crown-6 is a very poor catalyst in two-phase reactions. This generalization can hardly be justified by one observation, however, particularly since the 18-crown-6-catalysed reaction apparently leads to an equilibrium.

Ohtomi *et al.* (1976) have studied the catalytic effect of polypode ligands, such as [116] on the reactions of alkyl halides under liquid–liquid phase-transfer conditions (Table 34). Primary alkyl iodides are seen to be more reactive than the corresponding bromides. In contrast, the reactivity towards CN^- declines in the order RBr > RI > RCl. It is interesting to note that this order differs from that observed in solid–liquid two-phase systems catalysed by crown ethers (Cook *et al.*, 1974).

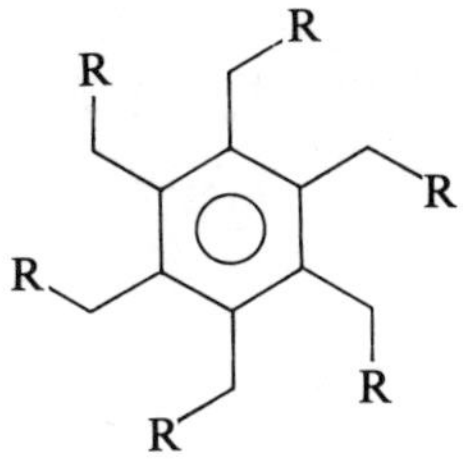

[116], $R = O(CH_2CH_2O)_3CH_3$

TABLE 34

Nucleophilic substitutions in alkyl halides catalysed by polypode [116] under liquid–liquid phase-transfer conditions[d]

Substrate	Reagent	Temp. /°C	Time /h	Yield[c] /%
$n\text{-}C_3H_7Cl$	KCN[a]	25	3	5 (0)
$n\text{-}C_3H_7Br$	KCN[a]	25	3	94 (2)
	KCl[b]	80	6	41 (5)
	KI[b]	80	3	94 (44)
$n\text{-}C_3H_7I$	KCN[a]	25	3	50 (1)
	KCl[b]	80	6	77 (4)
	KBr[b]	80	3	50 (2)

[a] 0.1 Molar equivalent of catalyst
[b] 0.05 Molar equivalent of catalyst
[c] Values in brackets give yields in the absence of catalyst
[d] Ohtomi *et al.*, 1976

Increasing the hydrophobicity of quaternary ammonium salts increases the apparent extraction constants for the ion pair and therefore leads to a higher catalytic activity (Brändström, 1977). The same phenomenon has been observed by Cinquini and Tundo (1976) for crown ether catalysis (Table 35). The catalytic activity of 18-crown-6 [3] and alkyl-substituted derivatives [117]–[119] in the reaction of $n\text{-}C_8H_{17}Br$ with aqueous KI follows the order: [117], [118] > [119] ≫ [3]. The alkyl-substituted [2.2]-cryptand derivatives are also much more efficient than the parent compound [86]. Increasing the hydrophobicity of [2.2.2]-cryptand (Cinquini *et al.*, 1975) and even of polypode ligands (Fornasier *et al.*, 1976) leads to higher catalytic activity. The tetradecyl-substituted compounds show the reactivity sequence: [2.2.2]-cryptand ≈ 18-crown-6 > [2.2]-cryptand on the reactivity scale that can be distilled from Table 35.

TABLE 35

Effect of catalyst hydrophobicity on the reaction of n-$C_8H_{17}Br$ and KI under liquid–liquid phase-transfer conditions at 60°C

Catalyst	Molar %	Time /h	Yield /%	Ref.
[118]	(5)	0.8	100	*a*
[119]	(5)	1.5	97	*a*
[120]	(5)	2	89	*a*
[121]	(5)	2	89	*a*
[122]	(5)	0.2	100	*b*
[123]	(5)	0.5	92	*b*
[37]	(5)	14	90	*b*
n-$C_{16}H_{33}P^+Bu_3Br^-$	(5)	1	93	*b*
[117]	(1)	5	98	*a*
[118]	(1)	3	90	*a*
[119]	(1)	5	71	*a*
[120]	(1)	5	50	*a*
[121]	(1)	5	60	*a*
[86]	(1)	5	~0	*a*
[124]	(1)	72	82	*c*
[125]	(1)	3	85	*c*
n-$C_{16}H_{33}P^+Bu_3Br^-$	(1)	3	72	*c*

[a] Cinquini and Tundo, 1976
[b] Cinquini *et al.*, 1975
[c] Fornasier *et al.*, 1976

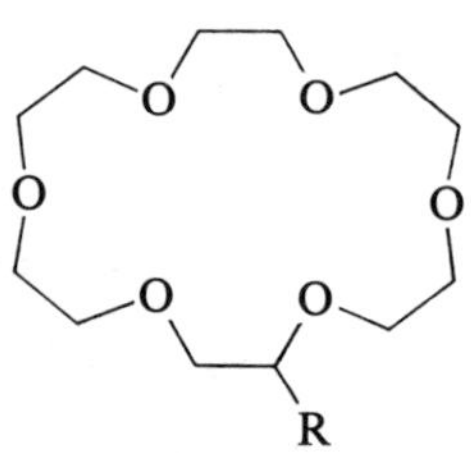

[117], R = n-$C_{20}H_{41}$

[118], R = n-$C_{14}H_{29}$

[119], R = n-$C_{10}H_{21}$

[3], R = H

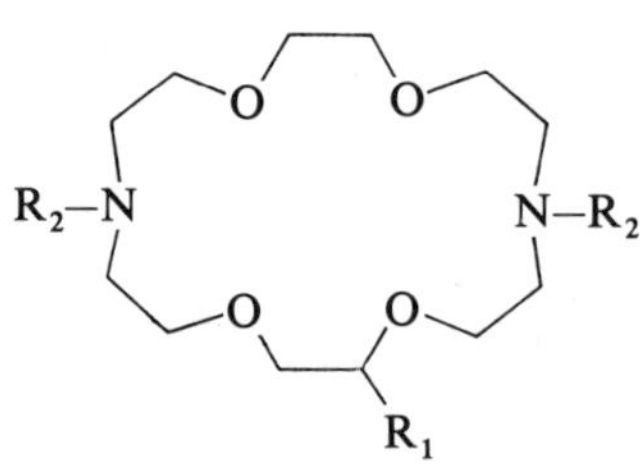

[120], R_1 = n-$C_{14}H_{29}$, R_2 = H

[121], R_1 = H, R_2 = n-$C_{16}H_{33}$

[86], $R_1 = R_2$ = H

Since polyethylene glycols and glymes of low molecular weight are very hydrophilic, it is not surprising that they are often less active than crown ethers in liquid–liquid systems (Movsumzade *et al.*, 1976). High-molecular-weight

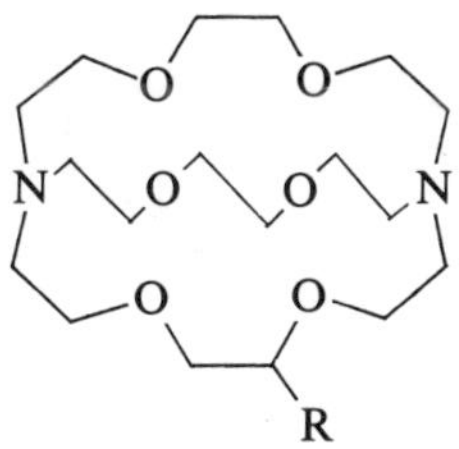

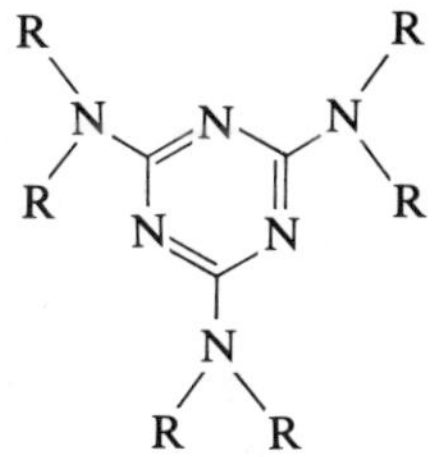

[122], R = n-$C_{14}H_{29}$ [124], R = n-$C_4H_9(OCH_2CH_2)_4$

[123], R = n-$C_{11}H_{23}$ [125], R = n-$C_8H_{17}(OCH_2CH_2)_4$

[37], R = H

(4×10^3) polyethylene oxide (PEO), however, was found to extract potassium salts of the anions SCN^-, I^-, Br^- into dichloromethane to almost the same degree as 18-crown-6 (Yanagida *et al.*, 1977c), and showed a higher activity than dibenzo-18-crown-6 in the reaction of 1-bromobutane with aqueous potassium iodide (Movsumzade *et al.*, 1976).

Herriott and Picker (1975) have studied the reaction between sodium thiophenoxide and 1-bromobutane in benzene–water catalysed by various quaternary ammonium salts and by the dicyclohexyl-18-crown-6 isomers ([20] + [21]). The catalytic activities, as judged from the second-order rate constants, span a range of 10^4. The best catalyst appeared to be dicyclohexyl-18-crown-6, directly followed by tetrabutylphosphonium chloride and tetrabutylammonium iodide.

The kinetics of the esterification of potassium *p*-nitrobenzoate by benzyl bromide in dichloromethane–water catalysed by dicyclohexyl-18-crown-6 ([20] + [21]) has been studied by Wong (1978). At low catalyst concentrations (e.g. 5.0×10^{-3} M) he found that 22% of the product was formed by the S_N1 and 78% by the S_N2 mechanism. Higher catalyst concentrations increased the salt concentration in the organic phase, and the contribution of the S_N1 mechanism becomes negligibly small.

Three-phase reactions

With a view to producing catalysts that can easily be removed from reaction products, typical phase-transfer catalysts such as 'onium salts, crown ethers, and cryptands have been immobilized on polymer supports. The use of such catalysts in liquid–liquid and liquid–solid two-phase systems has been described as "triphase catalysis" (Regen, 1975, 1977). Cinquini *et al.* (1976) have compared the activities of catalysts consisting of ligands bound to chloromethylated polystyrene cross-linked with 2 or 4% divinylbenzene and having different densities of catalytic sites ([126], [127], [132]–[135]) in the

TABLE 36

Reaction of n-$C_8H_{17}Br$ dissolved in toluene (1/3, v/v) with aqueous potassium salts in the presence of immobilized catalysts at 90°C[c]

		KI		KCN	
Catalyst (mol. eq.)	% ring substitution[a]	Time /h	Conversion /%	Time /h	Conversion /%
[127] (0.1)	50	16	—	16	19
[126] (0.1)	6.6	1.5	90	1.5	68
[126] (0.1)	10	1.5	90	1	86
[126] (0.1)	50	2.5	73	—	—
[126] (0.1)	50[b]	16	—	—	—
[126] (0.01)	10	5.5	89	3.5	90
[132] (0.01)	5	5	⩾95	—	—
[132] (0.01)	10	3	⩾95	—	—
[132] (0.01)	20[b]	6	⩾95	—	—
[135] (0.01)	2.5	3.5	⩾95	—	—
[135] (0.01)	5.0	3	⩾95	—	—

[a] 2% Divinylbenzene
[b] 4% Divinylbenzene
[c] Cinquini *et al.*, 1976

reactions of n-bromooctane, dissolved in toluene, with several aqueous potassium salts. From the results (Table 36) the authors conclude that the

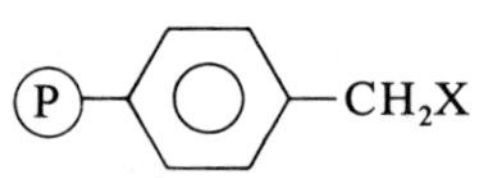

[126], X = $P^+(Bu)_3Cl^-$

[127], X = $N^+(CH_3)_3Cl^-$

[128], X = $N(CH_3)P(=O)[N(CH_3)_2]_2$

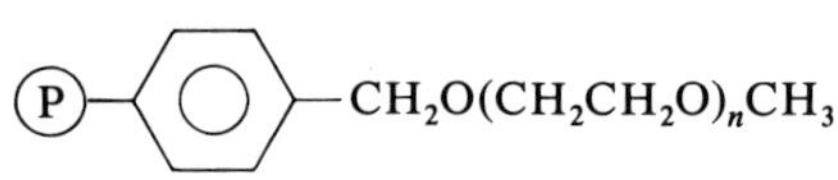

[129], $n = 1$

[130], n = 2

[131], $n = 3$

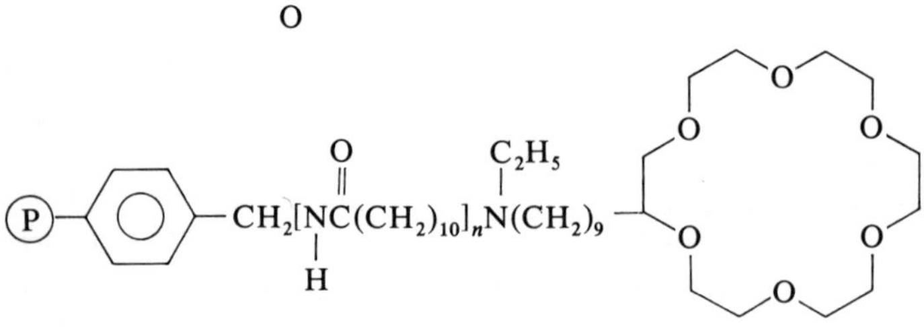

[132], $n = 0$

[133], $n = 1$

[134], $n = 2$

(P) = polymer support

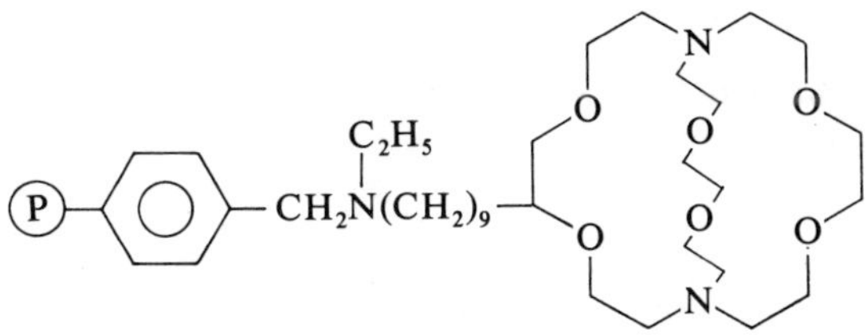

[135]

reaction rates are only slightly dependent on the number of catalytic sites along the polymer chain of the support. Furthermore, there seems to be no direct relationship between conversion times and concentration of the catalyst and, finally, the degree of cross-linking does not affect the catalytic activity of immobilized crown ether. Regen (1976), however, found that the rate of substitution of cyanide for bromide in 1-bromooctane under liquid–liquid–solid conditions was linearly dependent on the amount of catalyst (such as [127]) used.

The effect on the reaction rate of the length of the chain connecting the catalytic moiety with the polymer support (the so-called spacer) is shown in Table 37 (Molinari *et al.*, 1977). The activity clearly increases with increasing

TABLE 37

Half-life times of the reaction of n-$C_8H_{17}Br$ (in toluene) with aqueous potassium salts under triphase conditions[c]

Reagent	$t_{1/2}$/min					
	[126]	[132]	[133]	[134]	$C_{16}H_{33}P^+(Bu)_3$	[119]
KI[a]	180	53	45	42	36	35
KSPh[b]	240	70	45	39	20	33

[a] 90°C; 0.01 molar equivalent of catalyst
[b] 25°C; 0.01 molar equivalent of catalyst
[c] Molinari *et al.*, 1977

spacer length. With compound [134], the rate approaches that observed with n-decyl-18-crown-6 [119] under two-phase conditions. The effect of spacer length is solvent-dependent and is largest in highly apolar solvents such as n-heptane.

Triphase catalysis is applicable not only to liquid–liquid–solid (L, L, S) systems, but also to liquid–solid–solid (L, S, S) systems. For instance, McKenzie and Sherrington (1978) studied the substitution of phenoxide for bromide in 1-bromobutane using immobilized polyethyleneglycol monoethers such as [131] as triphase catalyst; they found (Table 38) that under L, L, S

TABLE 38

Reaction of 1-C_4H_9Br (0.48 mmol) in toluene (2 ml) with solid and aqueous phenoxide (2 mmol) at 25°C under triphase conditions[e]

Catalyst	(mmol)	Reagent	State	10^5k /M s^{-1}
None	—	NaOPh	Aqueous[d]	0.26
[131][a]	0.29	NaOPh	Aqueous[d]	0.56
	0.29	NaOPh	Anhydrous solid	3.2
	0.064	NaOPh	Anhydrous solid	2.5
	0.064	NaOPh	Solid from H_2O	1.4
	0.64	NaOPh	Anhydrous solid	8.3
	0.64	NaOPh	Solid from H_2O[c]	12
	0.29	KOPh	Anhydrous	11
	0.064	KOPh	Anhydrous	5.3
[131][b]	0.29	NaOPh	Anhydrous	3.2
	0.29	KOPh	Anhydrous	12

[a] 10% Cross-linked, 73% ring substitution
[b] 2% Cross-linked, 60% ring substitution
[c] Evaporated on immobilized catalyst
[d] 2 ml of water used
[e] McKenzie and Sherrington, 1978

conditions, the presence of [131] gives a twofold rate increase, but that under L, S, S conditions it produces an even more significant rate increase by a factor of *ca.* 6. The presence of small amounts of water apparently has a small effect. Potassium salts are about four times as reactive as sodium salts. It is interesting to note that under the conditions mentioned in Table 38, the initial rates are clearly dependent on catalyst concentration. Furthermore, the authors found immobilized catalysts to be *more* active than their non-immobilized counterparts.

Regen and Nigam (1978) recently found an interesting selectivity with respect to the size of the substrate in triphase catalysis. The relative rates of substitution of phenoxide for bromide in a homologous series of primary alkyl bromides catalysed by [126]–[130] are given in Table 39. It is seen that immobilized polyethyleneglycol monomethyl ethers show a large preference for the smaller bromides, in contrast to immobilized 'onium salts which exhibit only a slight size selectivity. From these and other data the authors concluded that the selectivity of catalysts [128], [129] and [130] is absorption-controlled whereas that of [126] and [127] is diffusion-controlled.

The detailed mechanism of triphase catalysed reactions still remains to be established. This holds in particular for reactions performed under liquid–

TABLE 39

Relative rates of reaction of alkyl bromides (in toluene) with aqueous sodium phenoxide under triphase conditions[c]

	Catalyst					
Substrate	[126][a]	[127][a]	[128][b]	[129][a]	[130][a]	$C_{16}H_{33}P(Bu)_3Br$
n-C_4H_9Br	1.40	2.07	5.29	62.00	64.00	1.06
n-$C_5H_{11}Br$	1.41	1.50	1.80	19.25	40.00	0.96
n-$C_6H_{13}Br$	1.17	1.44	1.75	5.09	6.89	1.02
n-$C_7H_{15}Br$	1.12	1.17	1.27	2.49	2.53	1.04
n-$C_8H_{17}Br$	1.00	1.00	1.00	1.00	1.00	1.00

[a] 17% Ring substitution
[b] 92% Ring substitution
[c] Regen and Nigam, 1978

solid–solid conditions with a combination of an immobilized catalyst (triphase catalyst) and a crown ether (phase-transfer catalyst), as was recently used by Regen *et al.* (1978) to convert benzyl chloride into benzyl cyanide.

NUCLEOPHILIC ADDITION

The earliest example of the increased reactivity of the potassium hydroxide complex of dicyclohexyl-18-crown-6 ([20] + [21]) was presented by Pedersen (1967b), who found that the sterically hindered esters of 2,4,6-trimethylbenzoic acid [136], which are inert towards KOH in protic solvents, are readily

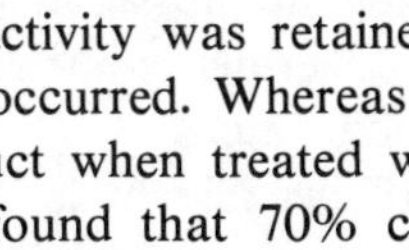

[136]

saponified by the complex (Table 40). Even the t-butyl and neopentyl esters react under these conditions. From the fact that in the saponification of the *l*-menthyl ester the optical activity was retained, the author concluded that normal acyl–oxygen fission occurred. Whereas the methyl ester in toluene at 74°C yielded 58% of product when treated with the crown-ether complex, Dietrich and Lehn (1973) found that 70% conversion was obtained after reaction for 12 h at 25°C with solid KOH and [2.2.2]-cryptand. These authors report that the reaction takes place not in a homogenous solution, as is the case with the crown ether complex, but at the solid–liquid interface.

An interesting decrease in rate was observed for the Cannizzaro reaction upon addition of 18-crown-6 [3] (Gokel *et al.*, 1976). With solid KOH and

TABLE 40

Saponification of esters of 2,4,6-trimethylbenzoic acid with potassium hydroxide complex of dicyclohexyl-18-crown-6[a,b]

Ester	Solvent	Concentration/M Ester	Concentration/M Base	Temp. /°C	Time /h	Conversion /%	k_2 /M^{-1} s^{-1}
Methyl	Toluene	0.122	0.122	74	31	58.4	1.08×10^{-4}
		0.120	0.120	100	2	53.0	1.17×10^{-3}
		0.061	0.154	105–111	5	93	—
	Benzene	0.061	0.154	80	5	39	—
t-Butyl	Benzene	0.060	0.154	80	5	22	—
	Toluene	0.060	0.154	104–111	5	94	—
Neopentyl	Benzene	0.060	0.154	80	5	40	—

[a] The complex is probably contaminated with small amounts of unsolvated hydroxide ions, methoxide ions, methanol and water

[b] Pedersen, 1967b

$$PhCHO + KOH \rightleftharpoons Ph{-}CH(OH)OK \qquad (15)$$

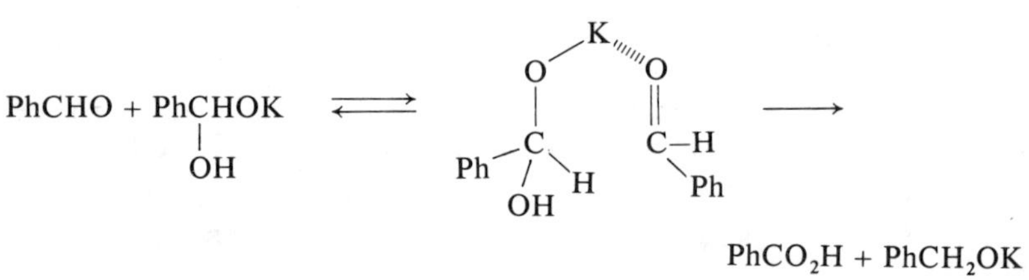

(16)

dichloromethane as the solvent, 24 h of stirring at room temperature gave an 89% yield of a mixture of equimolar amounts of acid and benzyl alcohol. In the presence of 1, 5 and 10 mol % of 18-crown-6 the yield of benzoic acid dropped to 58, 31, and 26% respectively. The authors rationalize this result on the basis of (15) and (16) by assuming that both the aldehyde and the alkoxide are co-ordinated to the metal cation in the six-membered transition state. The complexation of the cation by a crown ether reduces its Lewis acidity and hence the stability of the transition state.

The benzoin condensation of benzaldehyde in the presence of potassium cyanide has been studied by Akabori *et al.* (1976). Under liquid–liquid

$$2PhCHO \xrightarrow[\text{solvent}]{KCN} PhC(=O)CH(OH)Ph \qquad (17)$$

conditions (aqueous potassium cyanide; no solvent for the substrate) the benzoin was obtained in 3.1% yield at 25°C after 3.0 h; under the same conditions, the presence of 7 mol% 18-crown-6 increased the yield to 74%. Under liquid–solid conditions (solid KCN), the conversion was low in acetonitrile (1.1%) and benzene (0.1%) in the absence of 18-crown-6, but in the presence of the crown ether the yields were 78 and 71% respectively. The catalytic activity of dibenzo-18-crown-6 [11] is almost the same as that of 18-crown-6 [3].

Crown ether-assisted addition of halide to chloro-substituted ethylene-1,1-dicarboxylates [137] was reported to result in dechlorocarbomethoxylation (18) (Ykman and Hall, 1975). The authors proposed a mechanism in

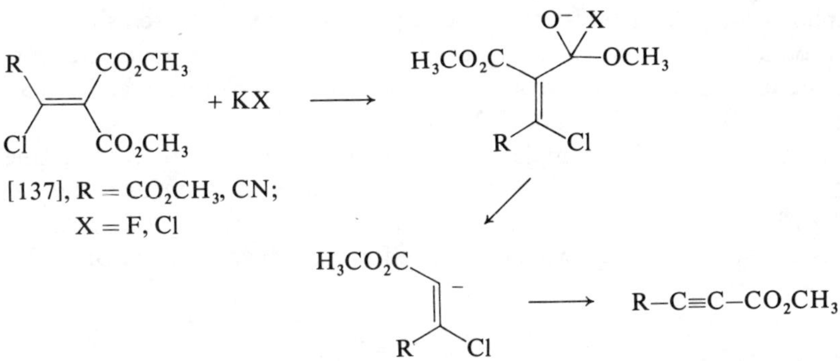

[137], R = CO_2CH_3, CN; X = F, Cl

(18)

which the addition of halide (solubilized by dicyclohexyl-18-crown-6 into sulfolane) is followed by elimination of a vinyl carbanion, which forms the acetylene. The strong nucleophilicity of the fluoride anion in potassium fluoride complexes of 18-crown-6 in benzene or acetonitrile is also apparent from the results of Yanagida *et al.* (1977a, 1977b). The 18-crown-6-catalysed addition of potassium carboxylates to perfluoro-2-methyl-2-pentene [138] gives the intermediate ester [139], which undergoes rapid nucleophilic fluoride addition

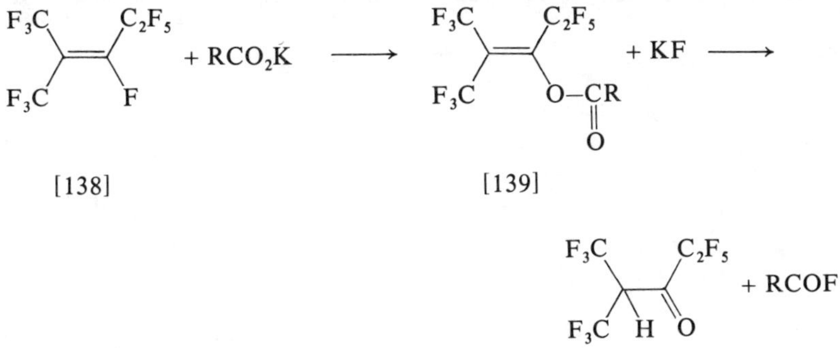

[138] [139]

(19)

to produce the acid fluoride after work-up (19). For reaction (19) heptaglyme was found to be as effective as 18-crown-6.

Yamamura and Murahashi (1977) have studied the crown ether-catalysed cyanation of vinyl halides under solid–liquid phase-transfer conditions (20). The reaction of *trans*-β-bromostyrene [140] with sodium cyanide in benzene,

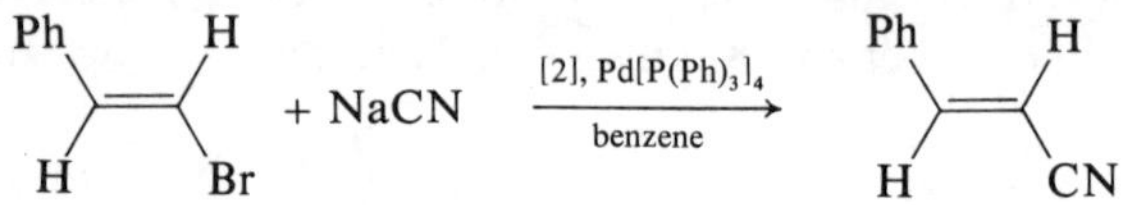

(20)

[140] [141]

catalysed by 15-crown-5 [2] (0.076 molar equivalents) and tetrakis(triphenylphosphine)palladium (0.03 molar equivalents) yielded exclusively the *trans* product [141]. Vinyl bromides were more reactive than the corresponding chlorides, and 18-crown-6 [3] was found to be more effective than either dibenzo-18-crown-6 [11] or [2.2.2]-cryptand [37].

The potassium cyanide complex of 18-crown-6 in benzene or acetonitrile undergoes Michael addition to α,β-unsaturated carbonyl compounds (Liotta *et al.*, 1977). In the presence of acetone cyanohydrin, the catalytic (i.e. catalytic in potassium cyanide and crown ether) cycle for hydrocyanation shown in (21)

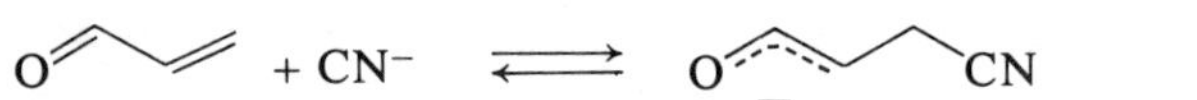

(21)

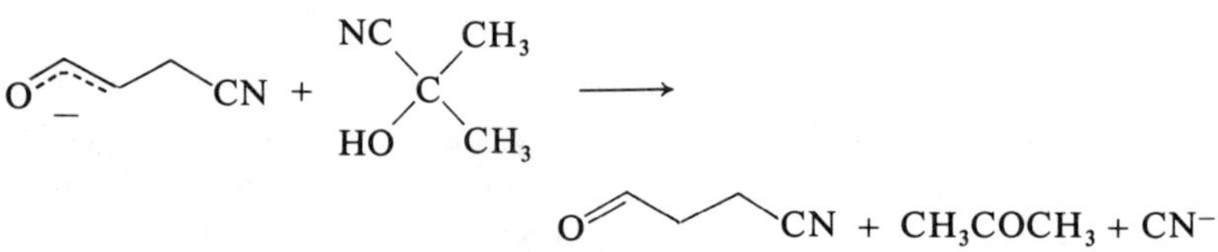

(22)

and (22) was accomplished. Application of this procedure to cholestenone gave high yields and good stereoselectivity.

Akabori and Yoshii (1978) used sodium hydride to generate the carbanions of α-chloro esters [142], which reacted with α,β-unsaturated esters to

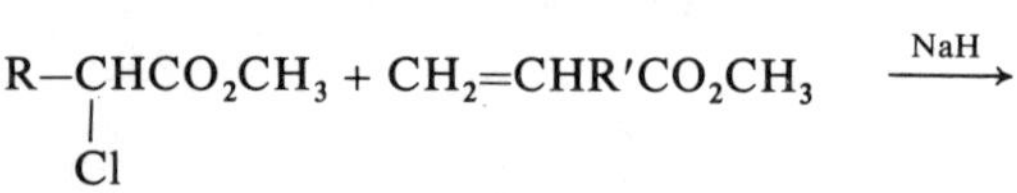

[142] [143]

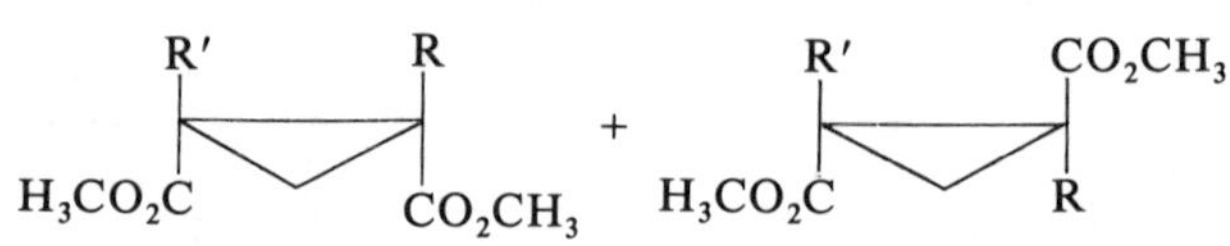

(23)

[144] [145]

TABLE 41

Yield (%) of *cis* [144] and *trans* [145] isomers in reaction (23) after 10 h[c]

R in [142]	R′ in [143]	Catalyst[a]	Benzene (20) [144]	Benzene (20) [145]	THF (25) [144]	THF (25) [145]	CH_2Cl_2 (20) [144]	CH_2Cl_2 (20) [145]	DMF (25) [144]	DMF (25) [145]	CH_3CN (20) [144]	CH_3CN (20) [145]
H	H	—	1.4	0.6	17.8	9.2	1.4	0.6	0	58	8	9
		[3]	6.9	14.1	8.5	16.5	8.0	60	0	14	23	49
		[37]	5.0	20.0	4.5	19.5	4.4	10.6	0	27	13	37
CH_3	CH_3	—	23.4	0.6	88.5	5.5	6.5	0.5	30	61	32	20
		[3]	42	38	26.4	33.6	29.0	55.0	26	58	16	38
		[3][b]	—	—	52.5	6.5	—	—	—	—	—	—
		[37]	23	35	7	19	7.0	16.0	7	18	8	30

[a] 0.08 Molar equivalents (based on NaH)
[b] In the presence of 8 molar equivalents of NaSCN
[c] Akabori and Yoshii, 1978

produce cyclopropane compounds (23). The product distribution as a function of solvent and catalyst is given in Table 41. In the absence of catalyst, the total yield and relative amount of *trans* product [145] increase with increasing cation-solvating properties of the solvent. Addition of either 18-crown-6 [3] or [2.2.2]-cryptand [37] increased the proportion of *trans* product, except in DMF where the product distribution is roughly constant. [2.2.2]-Cryptand [37] gives amost invariably a higher *trans*/*cis* ratio than 18-crown-6 [3]. The authors concluded from these results that the transition state leading to the *cis* isomer [146] is stabilized by cation chelation, whereas the free anion will adopt

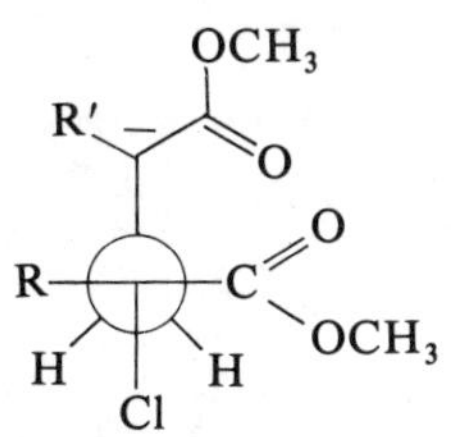

[146]

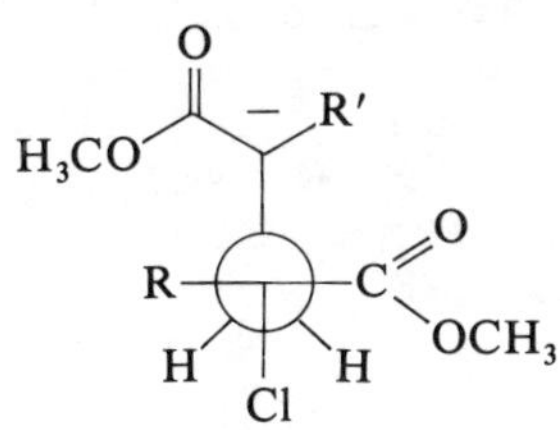

[147]

conformation [147] owing to steric repulsion and lead to the *trans* isomer. Addition of a cation-complexing agent will reduce the cation–anion interactions and consequently favour the formation of the *trans* product. This is confirmed by the fact that 18-crown-6 [3] shows no effect on the *trans*/*cis* ratio in the presence of a large excess of sodium cations (Table 41).

The effect of crown ether on the stereochemistry of the intramolecular cyclization of methyl (R)-6-α-methylbenzylaminocarbonylhex-2-enoate [148] has been studied by Wakabayashi and Kato (1977) in chlorobenzene solution. The anion was generated using catalytic (10%) amounts of both KOt-Bu and

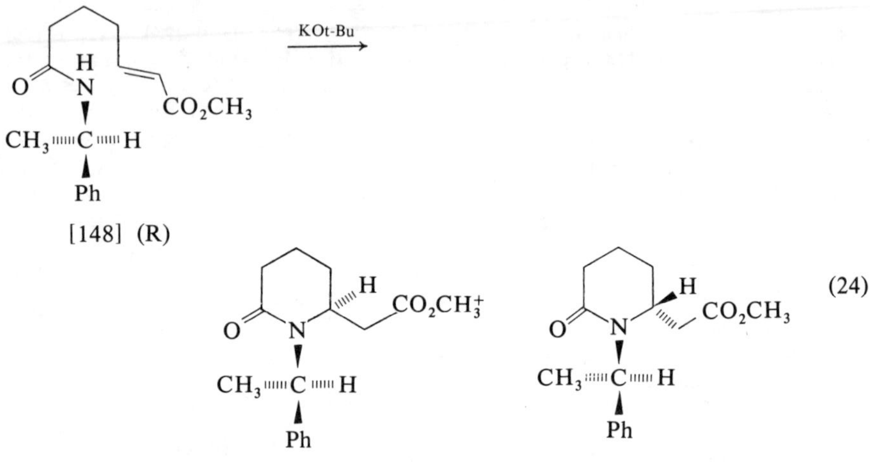

18-crown-6 [3]. In the absence of the crown ether, the diastereomeric esters [149] and [150] were produced in a ratio of 62 to 38 at −40°C (Table 42). In the presence of 18-crown-6 on the other hand, diastereomer [150] was

TABLE 42

Effect of 18-crown-6 [3] on the asymmetric cyclization of [148] with t-BuOK[a] in chlorobenzene[b]

Temp. /°C	Catalyst[a]	Time /min	Yield /%	Diastereomer excess /%
26	—	45	69	*S* (5.5)
	[3]	20	63	*R* (12.8)
−40	—	310	60	*S* (24.0)
	[3]	150	57	*R* (23.7)

[a] 0.1 Molar equivalent based on [148]
[b] Wakabayashi and Kato, 1977

favoured over [149]. The authors suggest that the reversed stereoselectivity results from the removal of the cation from the nucleophilic centre. The mechanism of this interesting diastereoselective reaction has not yet been clarified.

HYDROGEN-ABSTRACTION REACTIONS

The foregoing sections have dealt with the nucleophilic properties of anionic species whose salts were solubilized in apolar solvents with the aid of cation-complexing agents. Apart from the nucleophilic character of the anion, however, its basic strength is also expected to increase with decreasing extent of solvation. This section will deal with hydrogen-abstraction reactions of crown ether-solubilized bases.

Hydride

Buncel and Menon (1977) have studied reactions of potassium hydride solubilized in tetrahydrofuran by 18-crown-6 [3]. Very weak carbon acids such as triphenylmethane [151] ($pK_a = 31.5$), diphenylmethane [152] ($pK_a = 33.1$), and di-*p*-tolylmethane [153] ($pK_a = 35.1$) were completely metallated. After

Ph_3CH [151]

Ph_2CH_2 [152]

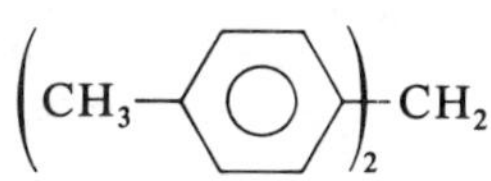

[153]

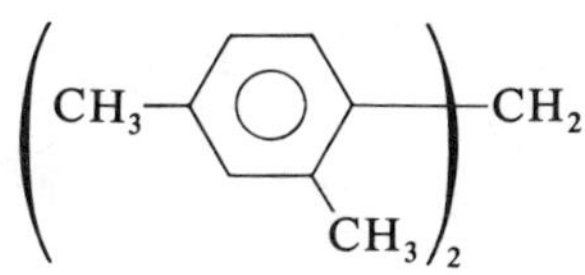

[154]

reaction for 72 h di-2,4-xylylmethane [154] ($pK_a = 36.3$) was only partially (20%) converted to the anion, while *p*-phenyltoluene ($pK_a = 38.6$) had not reacted at all under these conditions. From these results the authors concluded that compound [154] and H_2 are of comparable acidity. Although the amount of solubilized KH was very small ($21.5 \pm 9.5 \times 10^{-6}$ M), no metallation occurred in the absence of crown ether. The strongly basic properties of 18-crown-6-solubilized KH in THF have been used by Nordlander *et al.* (1978) for the N-alkylation of trifluoromethylacetamides.

Halide anions

Nucleophilic substitution reactions of halide anions in aprotic solvents are often accompanied by elimination reactions. For instance, reactions of secondary alkyl halides with potassium fluoride solubilized in acetonitrile with the aid of 18-crown-6 [3] give olefins as the main reaction product (Liotta and Harris, 1974). Similarly, the dicyclohexyl-18-crown-6 complex of potassium iodide acted exclusively as a base in its reaction with 2-bromo-octane in DMF (Sam and Simmons, 1974). The strongly basic character of weakly solvated fluoride has been exploited in peptide synthesis (Klausner and Chorev, 1977; Chorev and Klausner, 1976). It was shown that potassium fluoride solubilized in acetonitrile with 18-crown-6 [3] was capable of abstracting a proton from the nitrogen atom in indole. The same system (KF/18-crown-6/CH_3CN) was utilized by Belsky (1977) to generate carbanions from CH_3NO_2, EtO_2CCH_2CN, $CH_2(CN)_2$ for use in Michael additions.

The elimination of HBr from *cis*-*β*-bromo-*p*-nitrostyrene [155] producing the acetylene [156], has been studied by Naso and Ronzini (1974) using KF as the base (25). Their results (Table 43) show that the rate enhancement due to the

TABLE 43

Fluoride-promoted elimination of HBr from *cis*-*β*-bromo-*p*-nitrostyrene [155] under solid–liquid conditions[b]

Solvent	Temp. /°C	Catalyst[a]	Time /min	Conversion /%
CH_3CN	80	—	60	0
		([20] + [21])	60	53
DMF	70	—	30	28
		—	45	35
		([20] + [21])	30	58
		([20] + [21])	45	68
n-Butyl-cellosolve	150	—	10	15
		—	20	22
		([20] + [21])	10	43
		([20] + [21])	20	80

[a] Concentration of catalyst and substrate: 3×10^{-2} M
[b] Naso and Ronzini, 1974

presence of dicyclohexyl-18-crown-6 ([20] + [21]) is highest in acetonitrile, the solvent with the poorest cation-solvating properties.

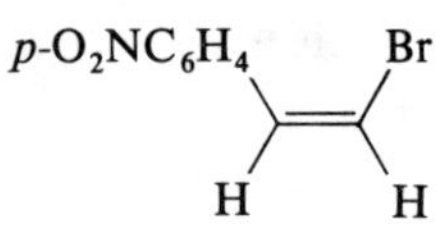

$$+ F^- \longrightarrow p\text{-}O_2NC_6H_4C{\equiv}CH \qquad (25)$$

[155] [156]

The conversion of benzaldoxime [157] into benzonitrile and benzamide by catalytic amounts of potassium halides and cyanide has also been attributed to proton abstraction by the anion (Rasmussen, 1977). Nucleophilic addition was thought to be unlikely, since the anion reactivities in (26) in acetonitrile with

$$2PhCH{=}NOH \xrightarrow[X = I, Br, CN]{X^-} PhC{\equiv}N + PhCONH_2 \qquad (26)$$

[157]

18-crown-6 as the catalyst decreased in the order: $CN^- \approx I^- \approx Br^- > Cl^- > F^-, OAc^-, OH^-$, which is different from the nucleophilicity scale in acetonitrile determined by Liotta and Grisdale (1975). It should be pointed out, however, that Liotta's nucleophilicity scale is based on differences in reactivity in homogeneous solution and that the overall rate of reaction in liquid–solid systems also depends on differences in solubility.

Carbonate, acetate, and cyanide ions

The solubilization of sodium or potassium carbonate into apolar solvents such as benzene or acetonitrile with the aid of 18-crown-6 [3] generates a powerful base that has been used for a variety of preparative transformations (Fedorynski *et al.*, 1978). Mechanistic studies have not been reported thus far.

Substitution reactions of cyanide with secondary alkyl halides are often accompanied by the formation of elimination products in variable amounts (Cook *et al.*, 1974). The same holds for reactions of metal acetate complexes of crown ethers (Liotta *et al.*, 1974).

Superoxide anion

Reactions of potassium superoxide solubilized in apolar solvents with crown ethers (see Oxidation reactions, p. 356) are also frequently accompanied by elimination reactions. Thus, in DMSO solution, secondary alkyl bromides only yield olefins when treated with the KO_2 complex of dicyclohexyl-18-crown-6 (Johnson *et al.*, 1978). Scully and Davis (1978) have studied the elimination of HCl from N-chloramines with 18-crown-6-solubilized KO_2, KOH, and KOAc in ether solution (27). High yields of aldimines were obtained with KO_2,

$$R_2CH\underset{\substack{|\\ Cl}}{N}R' + KX \xrightarrow[\text{18-crown-6}]{} R_2C{=}NR' \qquad (27)$$

whereas KOAc was unreactive. Although superoxide and acetate are of comparable basicity in water, superoxide is apparently a much stronger base in apolar solution. Oxidations by KO_2 initiated by hydrogen abstraction are discussed in the section on oxidation reactions.

Hydroxide ion

The strongly basic properties of potassium hydroxide are apparent from the work of Dietrich and Lehn (1973) who reported that the liquid–solid system KOH/THF/[2.2.2]-cryptand was capable of generating the anions of weak carbon acids such as triphenylmethane [142], diphenylmethane [143], and fluorene. The same anions could be generated using $NaNH_2$ instead of KOH.

Knipe *et al.* (1976) reported that N-(2-hydroxyethyl)-*p*-nitrobenzenesulphonamides [158] undergo a desulphonative double Smiles rearrangement in aqueous alkali (28) to produce the aniline derivative [160]. Upon replacing

$$\underset{[158]}{p\text{-}O_2NC_6H_4SO_2N(R)CH_2CH_2OH} \xrightarrow{OH^-} p\text{-}O_2NC_6H_4OCH_2CH_2N(R)\text{-}SO_2^- \longrightarrow \underset{[159]}{p\text{-}O_2NC_6H_4OCH_2CH_2NHR} \xrightarrow{OH^-} \underset{[160]}{p\text{-}O_2NC_6H_4N(R)CH_2CH_2OH} \quad (28)$$

the aqueous KOH by the solid–liquid system KOH/CH_2Cl_2/18-crown-6 the first Smiles rearrangement increased in rate relative to the second and the intermediate [159] could be isolated.

The effect of catalysts on the stereochemistry of the Horner–Wittig reaction under liquid–liquid conditions has been studied by Mikolajczyk *et al.* (1975). The *E*/*Z* isomer ratios in the product [162] proved to depend on the catalyst

$$\underset{[161]}{(EtO)_2P(=O)CH_2SPh} + PhCHO \xrightarrow[CH_2Cl_2/H_2O]{NaOH} \underset{[162]}{PhS\text{-}CH{=}CH\text{-}Ph} \quad (29)$$

used: $(n\text{-}Bu)_4N^+Cl^-$, 0.75; $(n\text{-}Bu)_4N^+Br^-$, 4.26; $(n\text{-}Bu)_4N^+I^-$, 4.88; dibenzo-18-crown-6 [11], 2.33; dicyclohexyl-18-crown-6 ([20] + [21]), 2.33. An explanation for this difference was not given.

For synthetic purposes, crown ethers have been used frequently as phase-transfer catalysts under conditions where the sodium or potassium hydroxide is present as a concentrated aqueous solution. Various anions of C—H acids have been generated this way (Makosza and Ludwikow, 1974). The use of such conditions for the generation of carbenes will be dicussed in a separate section.

Alkoxide ion

Conversion of tight ion pairs into crown ether-separated ion pairs leads in many cases to increased basicity. For example, Dietrich and Lehn (1973) have shown that a homogeneous solution of sodium t-amyloxide in benzene is unable to deprotonate triphenylmethane, whereas the reaction occurs rapidly in the presence of [2.2.2]-cryptand [37]. In THF or diethyl ether, alkali metal enolates do not react with triphenyl- or diphenylmethane (Pierre *et al.*,

1978a,b). In the presence of cryptands, however, the carbanions were formed instantaneously. These cryptated metal enolates were such strong bases that they even deprotonated THF and ether.

Roitman and Cram (1971) have shown that the kinetic basicity of potassium alkoxide increases on complexation with dicyclohexyl-18-crown-6 ([20] + [21]). Both the rate of isotopic exchange and the rate of racemization of (−)-4-biphenylylmethoxydeuteriomethane [163] by t-BuOK in t-BuOH were found to increase by factors between 30 and 17 000. Isotopic exchange was

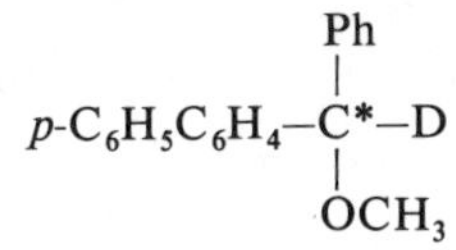

[163]

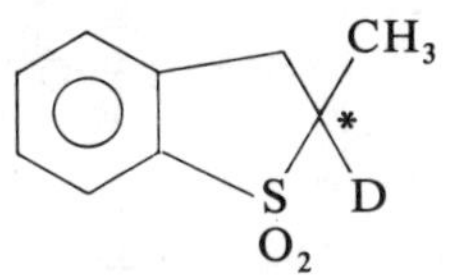

[164]

much less enhanced than racemization, however, the relative rate decreasing from 46 in the absence to about 1 in the presence of crown ether. The authors conclude that the degree of retention of configuration depends on the chelating capacities of the cation, which are greatly reduced on complexation with crown ether. Racemization and isotopic exchange of compound [164] by $KOCH_3$ in methanol were unaffected by the presence of dicyclohexyl-18-crown-6. In this system free anions or solvent-separated ion pairs are the prevailing basic species.

The rate of isomerization of 2-methylbicyclo[2.2.1]hepta-2,5-diene [165] to 5-methylenebicyclo[2.2.1]hept-2-ene [166] by t-BuOK in DMSO (30) was not affected by the presence of 18-crown-6 at low base concentrations (Maskornick, 1972). At high concentrations of t-BuOK (>0.1 M), the reaction is zero

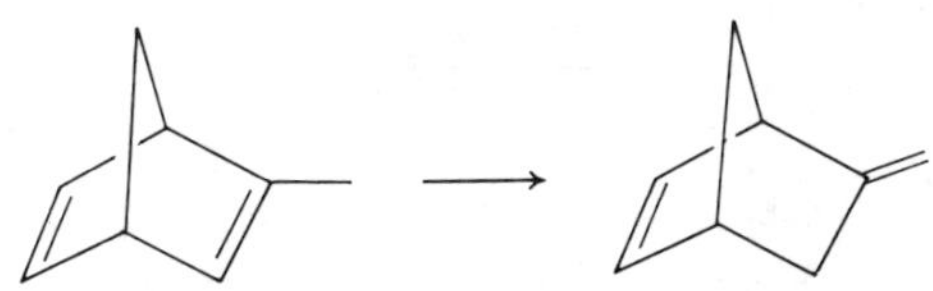

(30)

[165] [166]

order in base in the absence of 18-crown-6, and first order in the presence of 18-crown-6. In this case, the rate enhancement was ascribed to deaggregation of KOt-Bu ion pairs on complexation with crown ether.

The disproportionation of 1,3-cyclohexadiene to benzene and cyclohexene in dioxan is catalysed by KOt-Bu (El-Kholy *et al.*, 1977) according to reactions (31)–(33). The hydride transfer step (32) was found to be rate-determining. The presence of 18-crown-6 increased the overall rate of disproportionation by

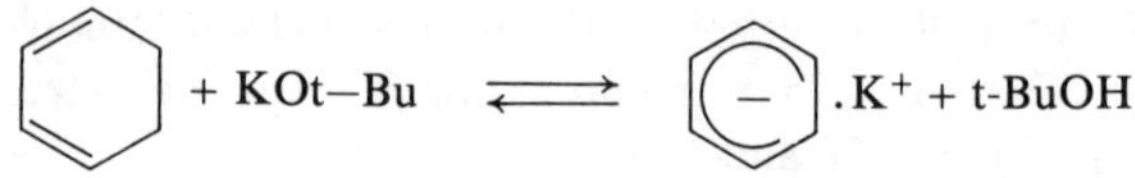 (31)

[167] [168]

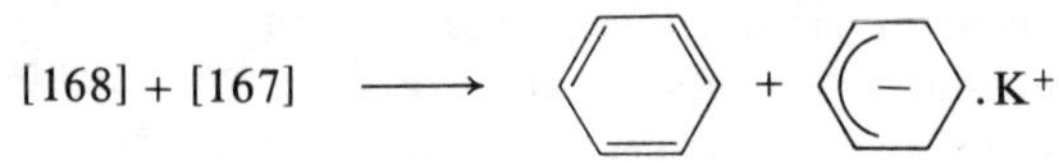 (32)

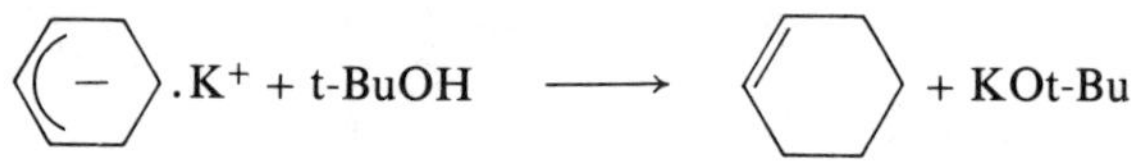 (33)

only a factor of 5. The authors conclude that hydride transfer is little affected by complexation of the cation and that the small rate-increase is due to a shift of equilibrium (31) to the right.

Since elimination and substitution reactions often compete with one another, Zavada and Pankova (1978) have studied the effect of dicyclo-18-crown-6 ([20] + [21]) on the E2/S_N2 ratio in the reaction of 1-nonyl tosylate and halide with potassium t-butoxide. In powerful cation-solvating solvents such as DMF and DMSO the free anion is the prevailing species in dilute solution, and consequently the presence of a crown ether does not change the E2/S_N2 ratio (Table 44). In poor cation-solvating solvents such as t-BuOH and benzene, on the other hand, the crown ether is seen to increase the E2/S_N2 ratio. The magnitude of this effect depends on the type of leaving group, the order being

TABLE 44

Effect of dicyclohexyl-18-crown-6 on the E2/S_N2 ratios in the reaction of n-C_9H_{19}X with t-BuOK[b]

Solvent	Temp. /°C	Time /min	Catalyst[a]	E2/S_N2 Ratio X = OTs	X = Cl	X = Br	X = I
Benzene	90	120	—	0.11	2.7	4.0	15.7
			([20] + [21])	1.0	15.7	24.0	24.0
t-BuOH	90	240	—	0.02	0.41	0.79	2.3
			([20] + [21])	0.075	1.3	1.2	2.6
DMF	20	10	—	0.25	6.0	3.2	4.3
			([20] + [21])	0.25	6.0	3.2	4.3
DMSO	20	10	—	0.25	4.6	3.0	5.9
			([20] + [21])	0.25	4.6	3.0	6.0

[a] Equimolar amount on base
[b] Zavada and Pankova, 1978

OTs > Cl > Br > I. The authors suggest that this sequence is due to changes in the S_N2 component.

It is clear from Table 44 that the $E2/S_N2$ ratios observed for reactions of crown ether-separated KOt-Bu ion pairs will greatly depend on the type of substrate and solvent. Di Biase and Gokel (1978) have recently reported many examples of the use of this reagent either as a nucleophile [for example, in its reaction with benzyl chloride and in the reaction with isatoic anhydride (34)] or as a strong base [for example, in the basic oxidation of fluorene to 2-carboxybiphenyl (35)].

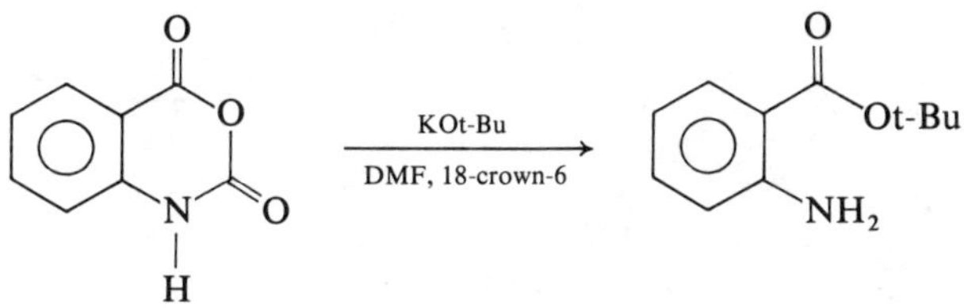

(34)

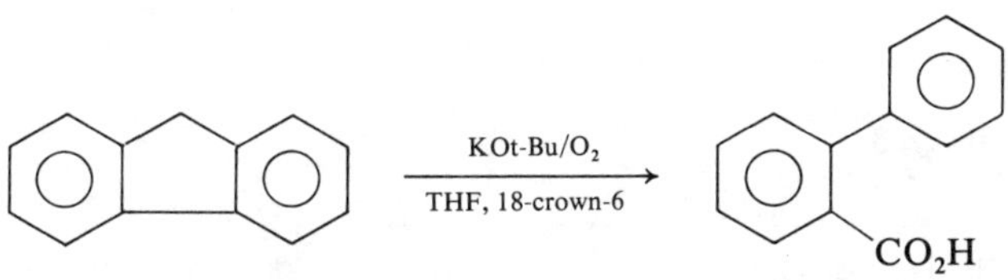

(35)

The effect of crown ethers on the geometrical orientation in base-promoted E2 eliminations has been studied by several groups. Bartsch *et al.* (1973) have investigated the effect of several parameters on the potassium alkoxide-promoted eliminations of HBr from 2-bromobutane (Table 45). In the absence of crown ethers the relative amount of 1-butene formed increases, while the *trans*/*cis* ratio of the 2-butenes decreases, with decreasing solvent polarity. Furthermore, the proportion of 1-butene increases and the *trans*/*cis* ratio decreases on increasing the base concentration. These effects were explained in terms of steric interactions between the base and α- and β-alkyl groups in the

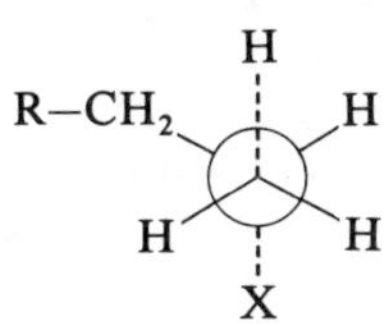

[169]

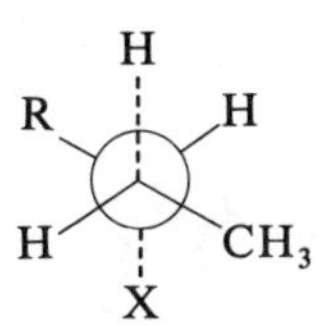

[170]

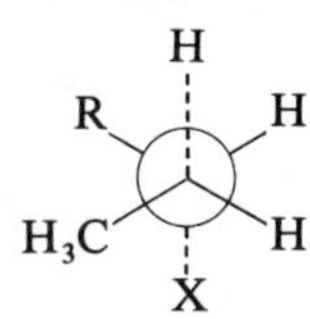

[171]

transition states for *anti*-elimination (Bartsch, 1975). For a bulky associated base steric interactions increase in the order [169] < [171] < [170], thus favouring 1-butene formation and *cis*-2-butene over *trans*-2-butene. On

TABLE 45

Effect of dicyclohexyl-18-crown-6 ([20] + [21]) on the potassium alkoxide (KOR)-promoted elimination of HBr from 2-bromobutane in ROH solution at 50°C[a,b]

ROH	[Base] /M	[Crown] /M	% 1-Butene	*trans*-2-Butene / *cis*-2-Butene
Methanol	0.25	—	15.4	3.34
	0.25	0.25	16.3	3.43
Ethanol	0.25	—	17.9	3.23
		0.25	17.6	3.31
Propan-2-ol	0.25	—	23.7	2.72
		0.25	23.8	2.95
t-Butyl alcohol	0.10	—	37.7	1.86
	0.25	—	41.6	1.78
	0.25	0.25	31.2	3.20
	0.50	—	44.1	1.66
	0.50	0.28	32.5	2.92
	1.00	—	49.9	1.48
3-Ethyl-3-	0.25	—	63.1	1.04
pentanol		0.25	37.3	2.82

[a] In DMSO solution, 0.25 M t-BuOK yields 30.3% 1-butene and a *trans*/*cis* ratio of 3.12
[b] Bartsch *et al.*, 1973

addition of crown ether, the complexation of the cation converts the bulky base aggregates into the much less sterically demanding crown ether-separated ion pair, thus creating conditions that are similar to those prevailing in eliminations conducted in DMSO (Table 45). A different way of preventing the formation of ion-pair aggregates was presented by Bartsch and Roberts (1977), who showed that self-solvating bases such as [172] and [173] gave small amounts of 1-butene and high *trans*/*cis* ratios of the 2-butenes in elimination reactions of 2-iodobutane in toluene solution.

$CH_3(OCH_2CH_2)_{7-8}OK$

[172]

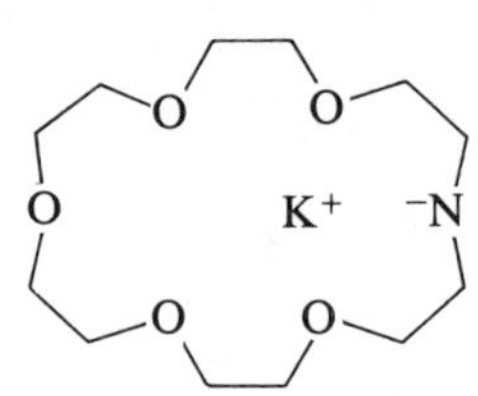

[173]

The changes in product distribution brought about by crown ethers depend not only on the base but also on the type of solvent, leaving group, and

TABLE 46

t-BuOK-promoted eliminations of HX from 2-$C_{10}H_{21}X$ in the absence (A) and in the presence (B) of equimolar amounts of dicyclohexyl-18-crown-6 ([20] + [21])[c]

		% 1-Decene		*trans*-2-Decene / *cis*-2-Decene	
Solvent	X	A	B	A	B
Benzene[a]	F	94	95.0	0.5	1.4
	OTs	88.2	81.8	0.7	1.7
	Cl	86.0	70.5	0.7	2.8
	Br	79.9	62.2	0.8	2.7
	I	66.1	46.5	1.5	3.2
t-BuOH[b]	F	92.4	91.4	0.8	2.9
	OTs	76.6	75.4	0.4	2.2
	Cl	84.5	64.6	1.1	3.6
	Br	79.4	53.3	1.3	5.1
	I	68.2	41.6	1.8	5.4

[a] Conditions: A, 130°C, 120 min; B, 20°C, 20 min
[b] Conditions A and B: 100°C, 120 min
[c] Pankova and Zavada, 1977

substrate. Several workers have observed that the crown-ether effect is very small in strongly cation-solvating solvents such as DMSO and DMF (Bartsch *et al.*, 1973; Pankova and Zavada, 1977; Zavada *et al.*, 1972). Pankova and Zavada (1977) have studied the effect of the leaving group. The results (Table 46) show that irrespective of the type of leaving group, crown ethers decrease the relative amount of 1-alkene produced and increase the *trans*/*cis* ratio of 2-alkenes. The magnitude of the crown-ether effect increases in the order of F $<$ OTs $<$ Cl $<$ Br $<$ I. These results were rationalized in terms of a variable transition-state theory (Zavada and Pankova, 1977). Base association has no effect on *anti*-eliminations from 'onium compounds (Bartsch, 1973; Zavada *et al.*, 1976). Consequently crown ethers showed no effect on the product distribution.

The structure of the substrate also plays an important role. The elimination reaction of compounds [174] in t-butyl alcohol promoted by KOt-Bu at 30°C

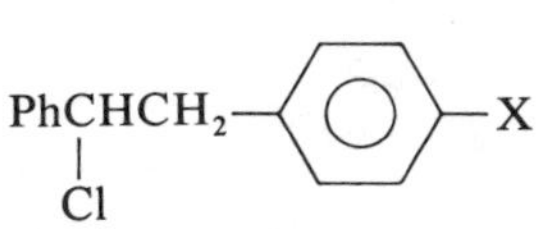

[174]

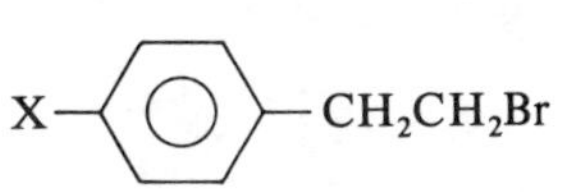

[175]

was characterized by the Hammett ρ-value of +2.20 in the absence and +3.40 in the presence of 18-crown-6 (Alunni *et al.*, 1977). The authors concluded that the carbanion character of the transition state increases on going from associated base to crown ether-complexed base. In contrast, the ρ-values found for 1-bromo-2-arylethanes [175] were +2.53 in the absence and +2.77 in the presence of 18-crown-6, indicating that, for this type of substrate, complexation of the base has little effect (Alunni *et al.*, 1976).

Zavada *et al.* (1973) have related the *trans/cis* olefin ratio in the product to competing *syn-* and *anti-*mechanisms for elimination. The transition state for *syn*-elimination [176] is stabilized by co-ordinative interactions with the cation.

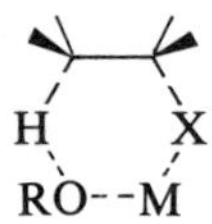

[176]

By saturating the co-ordination sites of the cation with a crown ether, the relative contribution of the *syn*-pathway decreases, as was shown by Zavada *et al.* (1972) for t-BuOK-promoted eliminations from 5-decyl tosylate. Furthermore, *trans/cis* ratios in both *anti-* and *syn*-eliminations were affected by the presence of dicyclohexyl-18-crown-6 and the nature of the leaving group (Zavada *et al.*, 1976). Fiandanese *et al.* (1973) observed that elimination from fluorosulphonylethanes [177] promoted by potassium phenoxide in dioxan

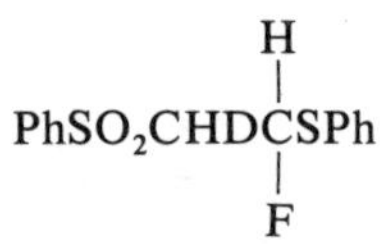

[177]

proceeds predominantly (87%) by the *syn*-pathway. In the presence of dicyclohexyl-18-crown-6 the importance of the *syn*-pathway was reduced to 42%.

Eliminations from cyclic compounds are more sensitive to the crown-ether effect. Potassium t-butoxide-promoted elimination of HX from [178] gives the *cis-* and *trans*-isomers of [179] and [180] (36), the *cis* isomers being formed

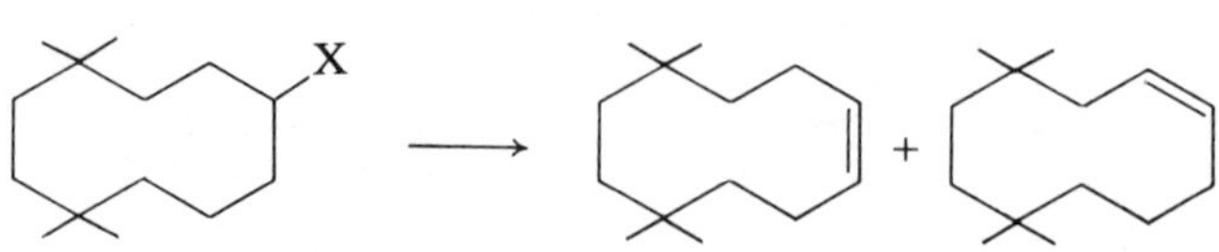 (36)

[178] [179] [180]

TABLE 47

t-BuOK-promoted eliminations of HX from cyclodecyl derivatives [178] in the absence (A) and in the presence (B) of dicyclohexyl-18-crown-6 ([20] + [21])[a]

			[179]			[180]		
X	Solvent		% *trans*	% *cis*	*trans*/*cis*	% *trans*	% *cis*	*trans*/*cis*
Br	benzene	A	83	1.5	55	14.0	1.5	9.3
		B	9.4	75.8	0.12	0.5	14.3	0.03
	t-BuOH	A	76.0	8.4	9.0	12.2	3.4	3.6
		B	8.9	75.3	0.12	1.1	14.7	0.07
	DMF	A	7.0	72.0	0.10	1.0	20.0	0.05
		B	3.7	79.4	0.05	0.1	16.8	0.006
OTs	benzene	A	88.8	1.4	63.4	8.8	1.0	8.8
		B	12.0	58.4	0.2	0.5	29.1	0.02
	t-BuOH	A	68.8	7.0	9.8	12.6	11.6	1.1
		B	22.2	54.6	0.4	2.8	20.4	0.14
	DMF	A	40.2	43.0	0.94	2.8	14.0	0.20
		B	5.8	68.7	0.08	0.3	25.2	0.01

[a] Svoboda *et al.*, 1972

mainly by *anti*-elimination, and the *trans* isomers by *syn*-elimination (Svoboda *et al.*, 1972). The results obtained in benzene (Table 47) show a dramatic shift from *syn*- to *anti*-elimination on addition of dicyclohexyl-18-crown-6 for both the bromide and the tosylate. Similar though somewhat smaller, effects were noticed for dicyclohexyl-18-crown-6 in the eliminations of HCl from chlorocyclodecane (Bartsch and Shelly, 1973). Further examples were provided by Bartsch and Kayser (1974), who found exclusively the *syn*-elimination product [182] when *exo*-2-norbornyl tosylate [181] was treated

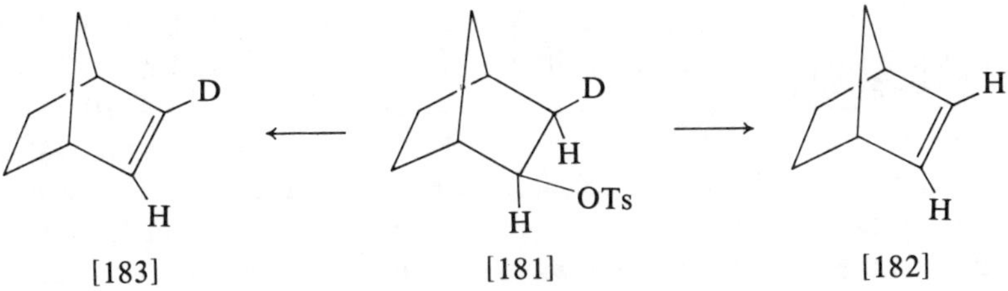

[183] [181] [182] (37)

with the sodium salt of 2-cyclohexylcyclohexanol in triglyme. In the presence of an equimolar amount of 18-crown-6, the *anti*-elimination product [183] was also formed in 27.2% yield (37). Decreasing *syn*-elimination in the presence of 18-crown-6 was also reported for the reaction of the compound [184] in the t-BuOK/t-BuOH system (Laureillard and Casadevall, 1977). The rate of elimination was increased by a factor of 265 by the addition of an equimolar amount of 18-crown-6. The rate enhancement can favourably be exploited in systems in which elimination and solvolysis compete. In ordinary base systems (for example, MeONa/MeOH) l-bornyl tosylate [185] only yields rearrange-

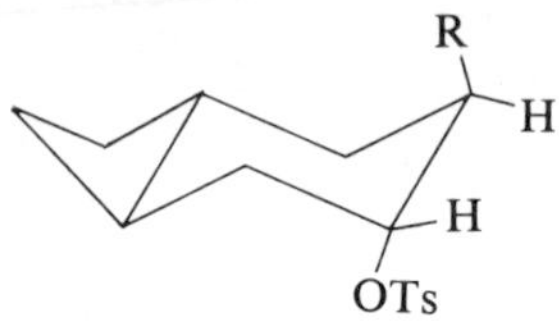

[184]

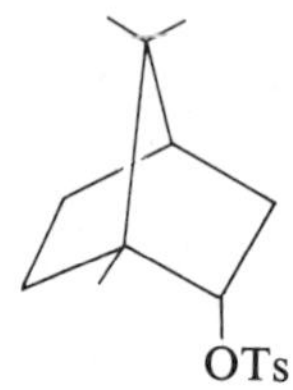

[185]

ment products derived from solvolytic elimination. The t-BuOK/benzene/18-crown-6 system only gives the E2-elimination product (Bartsch *et al.*, 1977). The presence of dicyclohexyl-18-crown-6 in the potassium t-butoxide promoted elimination of HX from *trans*-2-arylcyclopentyl tosylates [186]

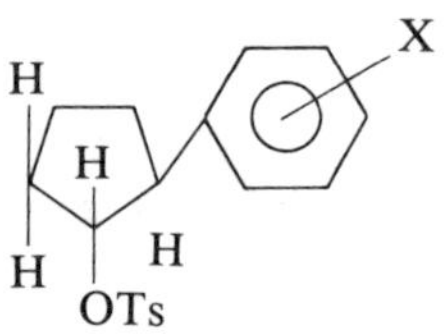

[186]

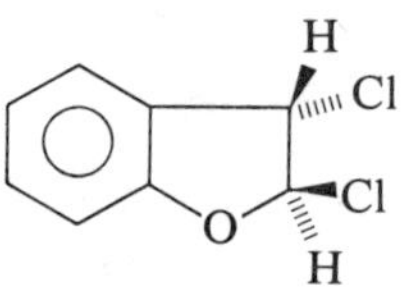

[187]

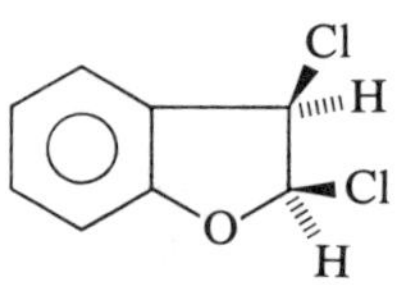

[188]

increased the relative amount of E2-*anti* product (3-phenylcyclopentene) and raised the Hammett ρ-value for E2-*syn* elimination from +2.2 to +3.1 (Bartsch *et al.*, 1974; Bartsch and Wiegers, 1972).

Baciocchi and Sebastiani (1979) have recently determined the rates of the elimination of, among other compounds, [187] and [188] promoted by KOt-Bu. The second-order rate constants ($M^{-1}\ s^{-1}$) at 30°C for [187] and [188] were 0.35 and 3.38×10^3 in the absence and 15.6 and $>2 \times 10^5$ in the presence of 18-crown-6. Hence, the rates of both *syn*- and *anti*-elimination increase on complexation of the cation, but the *anti*-pathway is much more favoured than the *syn*. The authors conclude that the detrimental effect of the crown ether on the co-ordinating power of the cation is largely outweighed by the increase in basicity of the anion.

Finally, crown ethers have also been shown to produce marked changes in the stereochemistry of E1cB-reactions of methoxyacenaphthenes [189] promoted by KOt-Bu/t-BuOH (Hunter and Shearing, 1971). Addition of dicyclohexyl-18-crown-6 increases the ratio of the rates of exchange and elimination from 1.3 to 3.2, while for both exchange and elimination the *cis*/*trans* ratio decreases.

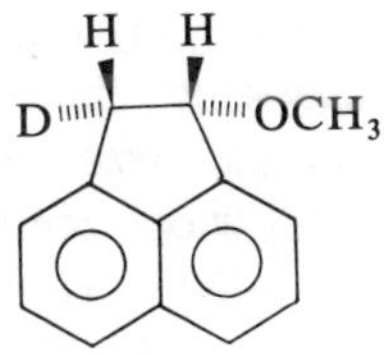

[189]

CARBENE GENERATION

Crown ethers have frequently been used as phase-transfer catalysts in the aqueous KOH/$CHCl_3$ liquid–liquid system to generate dichlorocarbene (Landini *et al.*, 1975; Makosza and Ludwikow, 1974; Sepp *et al.*, 1974; Kostikov and Molchanov, 1975). Assuming that product yields are a measure of the catalytic activity, Dehmlow and Lissel (1976) compared the activities of crown ethers and quaternary ammonium salts using cyclohexene as the carbene-intercepting substrate. The catalytic activities decreased in the order 18-crown-6 [3] > triethylbenzylammonium > dicyclohexyl-18-crown-6 ([20] + [21]) > dibenzo-18-crown-6 [11]. This sequence has no general validity, because under slightly different conditions or with other substrates the order can be reversed: dibenzo-18-crown-6 [11] > dicyclohexyl-18-crown-6 ([20] + [21]) (Landini *et al.*, 1975; Cinquini *et al.*, 1975); dibenzo-18-crown-6 [11] > triethylbenzylammonium (Kwon *et al.*, 1976). The exact origin of these changes is not yet known. Crown ethers are much better catalysts than 'onium compounds in the generation of mixed halocarbenes, such as :CBrCl, from $HCBr_2Cl$ (Fedorynski, 1977). Quaternary ammonium salts lead to the formation of a mixture of :CCl_2, :CBrCl, and :CBr_2, whereas dibenzo-18-crown-6 produces :CBrCl only. Dicyclohexyl-18-crown-6 ([20] + [21]) has been used to generate :CFCl from $HCCl_2F$ (Schlosser *et al.*, 1975). The generation of :CCl_2 from potassium trichloroacetate under solid–liquid conditions has been studied by Idemori *et al.* (1977). With cyclohexene as the substrate and benzene as the solvent, the relative activities decreased in the order dicyclohexyl-18-crown-6 > benzo-15-crown-5 > dibenzo-18-crown-6.

Sasaki *et al.* (1976) studied the phase-transfer catalysed generation of dimethylvinylidenecarbene [190] from 3-chloro-3-methyl-1-butene promoted

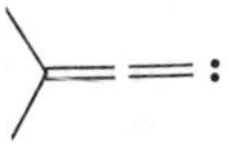

[190]

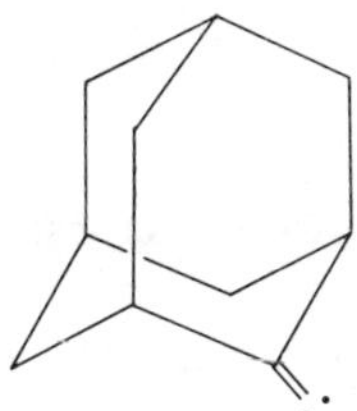

[191]

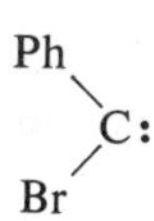

[192]

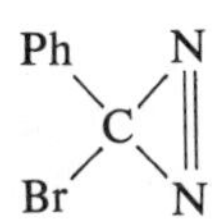

[193]

$PhCHBr_2$

[194]

by KOH under both liquid–liquid and solid–liquid conditions. They found crown ethers such as dibenzo-18-crown-6, dicyclohexyl-18-crown-6 and 18-crown-6 to be more effective catalysts than quaternary ammonium salts. The adamantylidenecarbene [191] was recently generated by α-elimination of HBr from ω-bromomethyleneadamante by t-BuOK in clyclohexene catalysed by 18-crown-6 (Sasaki *et al.*, 1978).

Moss and Pilkiewicz (1974) used crown ethers to determine the "freeness" of carbenic species by comparing the relative reactivities of various alkenes towards phenylbromocarbene [192], generated either by treatment of benzal bromide [194] with t-BuOK or by photolysis of phenylbromodiazirine [193]. The results (Table 48) show that the reactive species produced by these

TABLE 48

Relative reactivities of phenylbromocarbene [191] towards various alkenes[a]

Olefin	[194] + KOt-Bu	Precursor [194] + t-BuOK + 18-crown-6	[193]
Tetramethylethylene	1	1	1
Trimethylethylene	1.28	1.72	1.74
Isobutene	1.65	4.11	4.44
cis-But-2-ene	5.79	8.24	8.34
trans-But-2-ene	11.3	17.1	17.5

[a] Moss and Pilkiewicz, 1974

methods differ greatly in their reactivity towards alkenes. However, the species produced on adding one equivalent of 18-crown-6 to t-BuOK is identical with the photochemically generated free carbene. The authors conclude that the t-BuOK-generated species is a carbenoid—either a KBr or a t-BuOK complex—and that it is converted into the free carbene upon complexation of the cation by a crown ether. From the fact that the reactivity of $CH_3S(Cl)C$:, generated from CH_3SCHCl_2 by t-BuOK-induced α-elimination, was the same in the presence as in the absence of 18-crown-6, it was concluded that it was a free carbene (Moss *et al.*, 1975).

OXIDATION REACTIONS

Crown ethers have been used successfully as phase-transfer catalysts for liquid–liquid and liquid–solid oxidation reactions. Sam and Simmons (1972) observed that potassium permanganate can be solubilized in benzene by dicyclohexyl-18-crown-6 to yield concentrations as high as 0.06 M. From

^{55}Mn nmr spectroscopy the authors concluded that the crown ether-separated ion pair prevails in solution. At 25 °C, the complex reacts slowly (half life 48 h) with benzene. The complex reacts readily with olefins to give the potassium salts of the corresponding acids in high yields. The authors proposed a mechanism, in which the first step is a [2 + 4] electrocyclic addition as

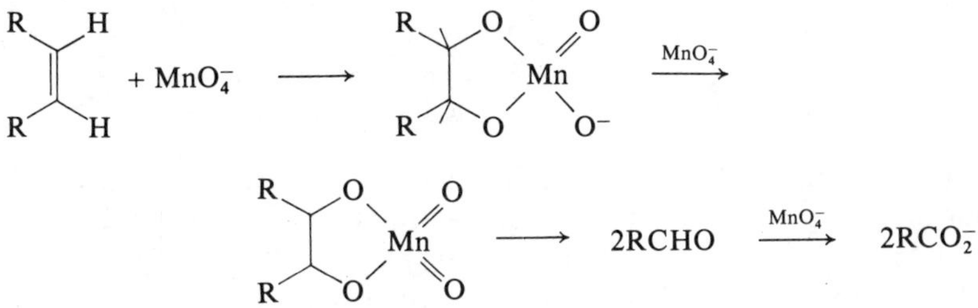

(38)

shown in (38). The same complex oxidizes alcohols to acids and toluene to benzoic acid. Similar results were reported by Landini *et al.* (1975). Since the manganese dioxide formed in the reaction rapidly covers the permanganate, mass transfer becomes a limiting factor, a problem which can be overcome by using ball-mill conditions (Pannell and McIntosh, 1977).

Lee and Chang (1978) have compared the ability of linear polyethers, crown ethers, and quaternary ammonium compounds to catalyse oxidations with $KMnO_4$ under liquid–liquid and liquid–solid conditions. In the presence of acetic acid as a scavenger for the KOH produced, the products of olefin oxidation were carboxylic acids, diones, diols and ketols. The three different classes of catalysts exhibited about the same activity in liquid–solid systems.

Cardillo *et al.* (1976) have used dicyclohexyl-18-crown-6 ([20] + [21]) and dibenzo-18-crown-6 [11] to solubilize K_2CrO_4 in hexamethylphosphoramide. The complex reacts with activated halides to produce aldehydes, according to (39).

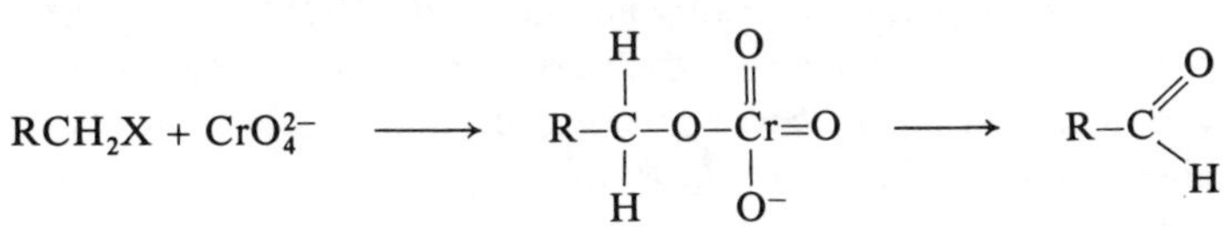

(39)

Valentine and Curtis (1975) extended the synthetic utility of potassium peroxide by reporting the successful solubilization of KO_2 in dry dimethyl sulfoxide using dicyclohexyl-18-crown-6 ([20] + [21]). Corey *et al.* (1975) used 18-crown-6 to solubilize KO_2 in dimethylformamide, dimethoxyethane and diethyl ether, whilst Johnson and Nidy (1975) reported its solubilization in benzene. A wide variety of chemical transformations have been realized with KO_2 complexes of crown ethers. With alkyl halides the main reaction products are peroxides, alcohols and olefins (Johnson and Nidy, 1975). Peroxides are formed by a mechanism in which the first step (40) (a nucleophilic displacement) is followed by an electron-transfer step (41) to form the alkylperoxy anion, which reacts with the substrate to form the product (42).

$$\text{R–X} + \text{O}_2^{-\bullet} \longrightarrow \text{R–O–O}^\bullet + \text{X}^- \qquad (40)$$

$$\text{R–O–O}^\bullet + \text{O}_2^{-\bullet} \longrightarrow \text{R–O–O}^- + \text{O}_2 \qquad (41)$$

$$\text{RX} + \text{R–O–O}^- \longrightarrow \text{R–O–O–R} + \text{X}^- \qquad (42)$$

Consistent with this mechanism is the fact that the reaction takes place with inversion of configuration at the reaction centre (Johnson and Nidy, 1975). For alkyl tosylates similar results have been reported (San Fillippo *et al.*, 1975). The formation of alcohols under these conditions has been ascribed to secondary reactions (Chern *et al.*, 1978; Johnson *et al.*, 1978). Under certain conditions alcohols are even the only isolated products (Corey *et al.*, 1975).

Moro-Oka *et al.* (1976) have reported that the oxidation of 9,10-dihydroanthracene by KO_2 solubilized in DMSO by 18-crown-6 gives mainly the dehydrogenated product, anthracene. Under the same conditions, 1,4-hexadiene is dehydrogenated to benzene. The authors proposed a mechanism in which the superoxide ion acts as a hydrogen-abstracting agent only. The oxidations of anthrone (to anthraquinone), fluorene (to fluorenone), xanthene (to xanthone) and diphenylmethane (to benzophenone) are also initiated by hydrogen abstraction.

Aromatic halides react with crown ether-complexed KO_2 by an electron-transfer mechanism and not by nucleophilic attack, as was shown by Frimer and Rosenthal (1976) using esr spectroscopy. The corresponding phenol is the main reaction product (Yamaguchi and Van der Plas, 1977). Esters are saponified by the KO_2/18-crown-6 complex in benzene, presumably by an addition-elimination pathway (San Fillippo *et al.*, 1976). The same complex has been used to cleave α-keto-, α-hydroxy-, and α-halo-ketones, -esters, and -carboxylic acids into the corresponding carboxylic acids in synthetically useful quantities (San Fillippo *et al.*, 1976).

REDUCTION REACTIONS

Pierre and Handel (1974) observed that cyclohexanone was cleanly reduced to cyclohexanol by $LiAlH_4$ in diglyme solution. On addition of one equivalent (based on Li^+) of the strong Li^+-complexing agent [2.1.1]-cryptand, all the $LiAlH_4$ was solubilized, but the reduction was completely inhibited. In the presence of additional amounts of either LiI or NaI, the ketone was reduced in the normal way. On the basis of these results, the authors concluded that the cation is indispensable for the reduction and they proposed a mechanism in which hydride transfer occurs via [195], in which the carbonyl moiety is co-ordinated to the cation. Reactions of $NaBH_4$ in diglyme were also completely inhibited by the presence of equimolar amounts of [2.2.1]-cryptand [36]. The fact that reduction did take place after hydrolytic work-up or in methanol solution, led the authors to propose a solvent-assisted hydride transfer via [196] in the case of BH_4^- reductions. Cation catalysis in metal

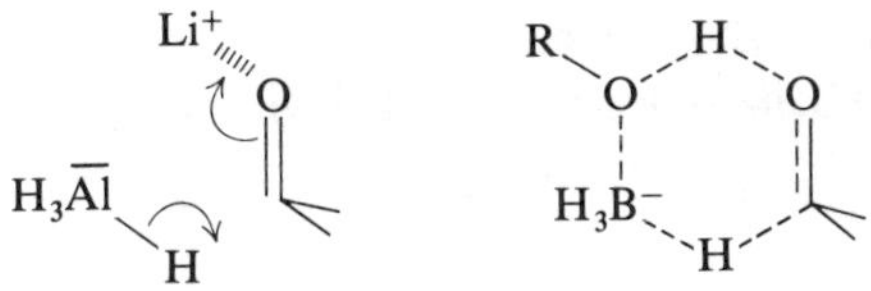

[195] [196]

hydride reductions was also observed for other ketones, aldehydes, carboxylic acids, amides, esters, nitriles, oximes, tosylates, and epoxides (Pierre *et al.*, 1975; Handel and Pierre, 1975b; 1976). Furthermore, both 1,2- and 1,4-additions to α,β-unsaturated ketones by $LiAlH_4$ or $NaBH_4$ were inhibited by the presence of equimolar amounts of cryptands (Handel and Pierre, 1975a).

The effect of cryptands on the reduction of ketones and aldehydes by metal hydrides has also been studied by Loupy *et al.* (1976). Their results showed that, whereas cryptating the lithium cation in $LiAlH_4$ completely inhibited the reduction of isobutyraldehyde, it merely reduced the rate of reduction of aromatic aldehydes and ketones. The authors rationalized the difference between the results obtained with aliphatic and aromatic compounds in terms of frontier orbital theory, which gave the following reactivity sequence: Li^+-co-ordinated aliphatic C=O $\approx$ Li^+-co-ordinated aromatic C=O $\gg$ non-co-ordinated aromatic C=O $>$ non-co-ordinated aliphatic C=O. By increasing the reaction time, Loupy and Seyden-Penne (1978) showed that cyclohexenone [197] was reduced by $LiAlH_4$ and $LiBH_4$, even in the presence of [2.1.1]-cryptand, albeit much more slowly. In diethyl ether in the absence of

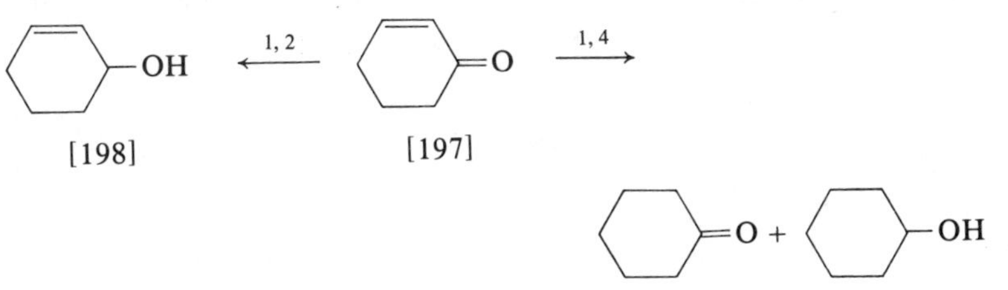

[198] [197] [199] [200] (43)

ligand, the ratio of ([199] + [200]) to [198] was 0.02 (almost exclusively 1,2-addition), whereas in the presence of [2.1.1]-cryptand, it was 3.34 (mainly 1,4-addition). This inversion of regioselectivity brought about by cation complexation was explained in terms of MO theory (Lefour and Loupy, 1978).

Crown ethers have also been found to lower the rate of reduction by metal hydrides. Wiegers and Smith (1978) reported that the rate of reduction of camphor by $LiAlH_4$ in tetrahydrofuran was depressed by a factor of 6 on addition of one equivalent of crown ether [201]. They also concluded that, although the free cation shows a catalytic effect in metal hydride reduction, it is not indispensable. Dibenzo-18-crown-6 [11] was also found to lower the rate of

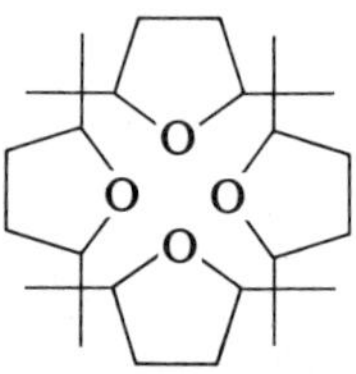

[201]

reduction of aldehydes and ketones by sodium borohydride in tetrahydrofuran (Hirao *et al.*, 1978c). In this case, however, the authors observed the formation of an insoluble 1 : 1 complex, and the lower rate of reduction might be due to precipitation of $NaBH_4$ from the solution. In apolar solvents such as toluene, ketones are reduced by $NaBH_4$ in the presence of dibenzo-18-crown-6 (Matsuda and Koida, 1973). Besides the alcohol, considerable amounts of products were formed arising from base-catalysed condensation of the ketone.

Dupont Durst *et al.* (1974) have used 18-crown-6 to solubilize $NaBH_3CN$ in a methanol/dichloromethane mixture, the solution reducing alkoxysulfonium salts to the corresponding sulfides in high yields.

MISCELLANEOUS REACTIONS

Hunter *et al.* (1974) have reported a striking catalytic effect of dibenzo-18-crown-6 [11] on the decarboxylation of sodium 3-(fluoren-9-ylidene)-2-phenylacrylate [202] in THF solution. At −45°C the salt failed to form a

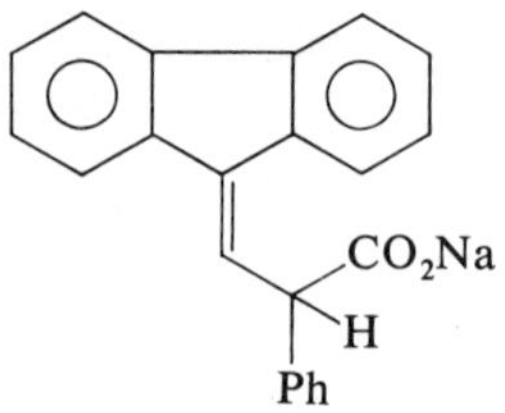

$Ph_3C{-}CO_2^-M^+$ ϕ-N=N−CO_2K

[202] [203] [204]

detectable amount of decarboxylated product even after 200 minutes. In the presence of 8 × 10^{-6} M crown ether, however, the reaction started immediately, indicating a rate enhancement by a factor of at least 10^5. The authors suggested that the decarboxylation of the complex might even be faster than complex formation. This interesting suggestion was not substantiated in more recent work on the decarboxylation of a variety of salts catalysed by 18-crown-6 (Hunter *et al.*, 1978). The rates of decarboxylation of lithium, sodium, and potassium triphenylacetates [203] were increased by a factor of about 10^2 by the presence of equimolar amounts of 18-crown-6. This factor is the resultant of the anion-activating and solubility-increasing properties of the crown ether. The mechanism of decarboxylation appeared to be rather complicated. In the presence of 2–5 moles of water per mole of the sodium salt [203], the observed rate constant showed a half-power dependence on crown-ether concentration, a finding which was rationalized by assuming a mechanism in which the free anion rather than the crown-ether complex is the reactive species. The ability of 18-crown-6 to accelerate both the hydrolysis of esters and the decarboxylation of metal carboxylates was exploited by Hunter and Perry (1977) in the malonic ester synthesis. Fraenkel and Pechhold (1970) used 18-crown-6 to solubilize potassium phenylazoformate [204] in THF and studies its decomposition into phenylpotassium, nitrogen and carbon dioxide. Crown ether-complexed phenylpotassium is not stable in THF and forms benzene by abstraction of a proton from the solvent.

Evans and Golob (1975) have studied the effect of 18-crown-6 on the rate of the oxy-Cope rearrangement (44). In THF solution at 0°C, the rate increased

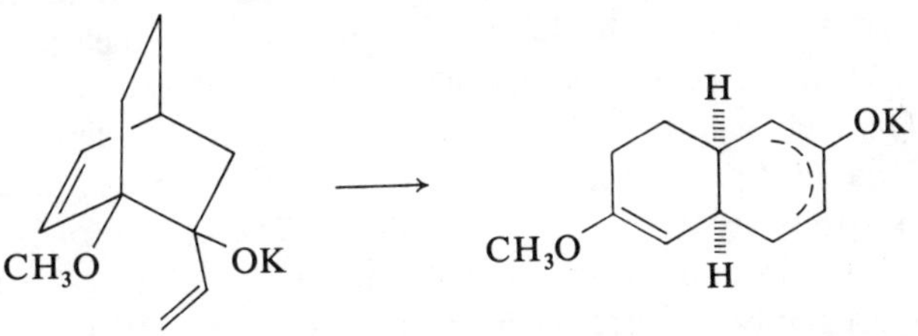

(44)

by a factor of 180 (plateau value at about 2 equivalents of 18-crown-6). With the corresponding sodium salt, 15-crown-5 caused only a 1.27-fold rate increase.

Gold and Sghibartz (1978) have studied the effect of cations on the acid-catalysed hydrolysis of crown ether acetals such as [205]. For small ring

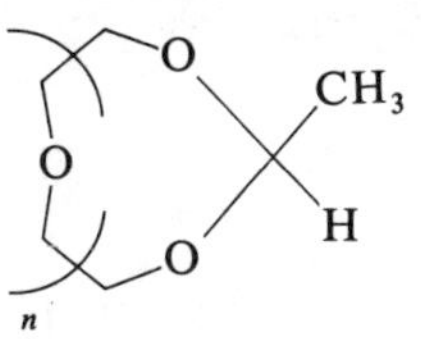

[205]

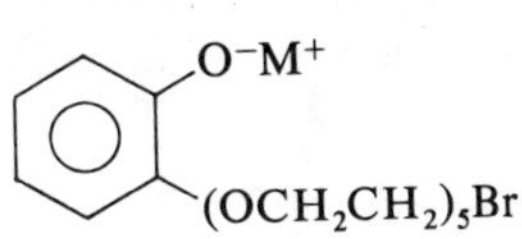

[206]

sizes ($n = 0$–3) all the alkali cations exhibited a small rate-enhancing effect (1.0–1.6 fold). In contrast, the rate of hydrolysis of 2-methyl-17-crown-6 ($n = 4$) was depressed by cations, the activity decreasing in the order $Na^+ > K^+ > Rb^+ > Cs^+$. With 2-methyl-20-crown-7 ($n = 5$) the sequence was $Rb^+ > K^+ > Cs^+ > Na^+$. The results are consistent with the assumption that the introduction of a cation into the cavity lowers the basicity of the acetal oxygen atoms.

Whereas Gold and Sghibartz showed that cation complexation depressed the rate of crown-ether disrupture, there is convincing evidence that crown ether formation is facilitated by the presence of cations. The template effect, presumably due to complexation of the open-chain precursor and formation of a crown-type conformation, clearly emerged from studies in which the yield of crown ethers was related to the type of cations present (Reinhoudt *et al.*, 1976). Kinetic evidence for the template effect was presented by Mandolini and Masci (1977), who showed that the rate of cyclization of the precursor of benzo-18-crown-6 [206] decreased in the order: $Ba^{2+} > Sr^{2+} \gg K^+ > Na^+ \gg Li^+$. This sequence is the same as the one found for the stability constants of the 1 : 1 complexes of these cations with 18-crown-6 in water (Table 3).

5 Complexes with protonated amines

GENERAL

The complexation of protonated amines, RNH_3^+, by crown ethers differs in many aspects from the complexation of metal cations. Whereas complexes with metal cations derive their binding energy mainly from electrostatic forces, complexes with ammonium ions are likely also to be stabilized by hydrogen bonding. The relative contributions of hydrogen bonding ($N^+{-}H\cdots O$ interaction) and electrostatic interactions ($N^+\cdots O$) were estimated to be 75% and 25%, respectively (Timko *et al.*, 1977). Therefore, for strong RNH_3^+ complexation, the location of the binding sites of the crown ether should allow a tripod hydrogen-bond arrangement. Furthermore, the low symmetry of RNH_3^+ cation compared to that of the spherical metal cations offers the possibility of distinguishing cations by their shape. The organic group R in the cation may also interact with the crown ether. These interactions can either destabilize the complex by steric repulsion or stabilize it by hydrophobic bonding, dipole–dipole and charge-transfer interactions. CPK molecular models are indispensable for establishing which interactions prevail in a particular combination of cation and crown ether.

THERMODYNAMIC STABILITY

The effect of the organic group R on the stability of RNH_3^+ complexes of 18-crown-6 in methanol has been determined by Izatt *et al.* (1978b,c) by means of a calorimetric titration technique. It was found that in this solvent the type of anion had no effect on the complex stability. The results (Table 49) show that the log K values of alkylammonium ions are subject to only minor variations, the association constants being about one hundredth of that for K^+.

TABLE 49

Log K, ΔH, and $T\Delta S$ values for RNH_3^+ complexes of 18-crown-6 in methanol at 25°C[a]

R	Log K	ΔH /kcal mol^{-1}	$T\Delta S$ /kcal mol^{-1}
H	4.27	−9.27	−3.5
Me	4.25	−10.71	−4.9
Et	3.99	−10.65	−5.2
n-Pr	3.97	−10.06	−4.6
i-Pr	3.56	−9.65	−4.8
Cyclopropyl	3.90	−10.12	−4.8
t-Bu	2.90	−7.76	−3.8
OH	3.99	−9.01	−3.6
NH_2	4.21	−10.43	−4.7
Ph	3.80	−9.54	−4.4
4-Methylphenyl	3.82	−9.92	−4.7
3,5-Dimethylphenyl	3.74	−9.07	−4.0
2-Methylphenyl	2.86	−7.59	−3.7
2,6-Dimethylphenyl	2.00	−5.65	−2.9

[a] Data from Izatt *et al.*, 1978b; 1978c

The t-butylammonium cation forms a relatively weak complex, which was attributed to steric repulsion. The stability of anilinium complexes is about the same as that of alkylammonium complexes. The introduction of substituents at the 2- and 6-positions increases the steric repulsion which results in a drop in binding constant. This effect is enthalpic in origin. These results show that variations in the group R have only a minor effect on the stability of complexes in which the crown ether has a fully exposed cavity. Furthermore, there is no relationship between the pK_a-value of RNH_3^+ and the stability of the complex.

Cram and co-workers (Kyba *et al.*, 1977b; Timko *et al.*, 1974; Newcomb *et al.*, 1977b; Timko *et al.*, 1977; Helgeson *et al.*, 1977) have studied many parameters that affect the binding of t-butylammonium cations by crown ethers. The association constants were determined from extraction experiments; some of the results are collected in Table 50. Crown ethers [207]–[219] constitute a series of compounds in which one, two, or three CH_2OCH_2 units are replaced by several other groups (A–F). The main results are as follows. (*i*) The macrocyclic effect for 18-crown-6 [3] is about 10^4. (*ii*) 18-Membered rings form more stable complexes with t-$BuNH_3SCN$ than do 12-, 15-, and

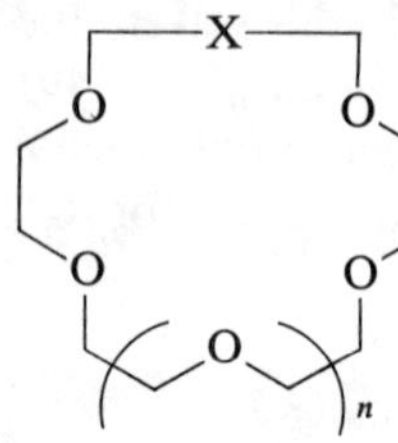

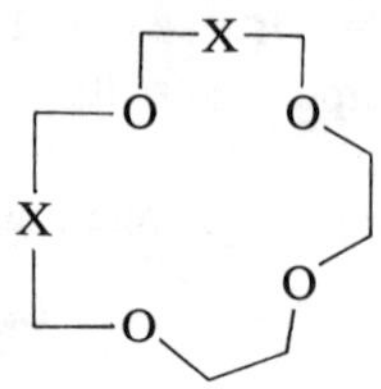

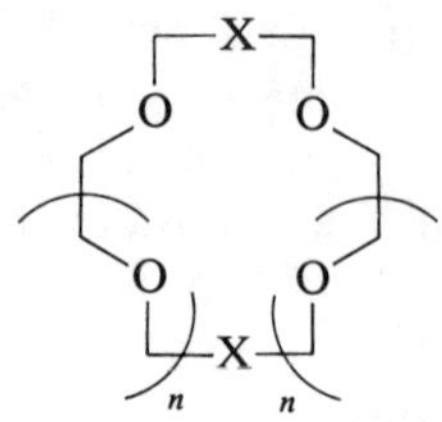

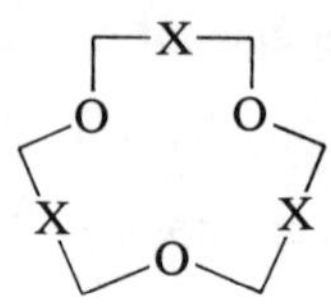

[207], X = A, $n = 1$; [3], X = B, $n = 1$; [28], X = C, $n = 1$; [208], X = D, $n = 1$; [209], X = E, $n = 0$; [210], X = E, $n = 1$; [79], X = F, $n = 1$

[211], X = D; [212], X = E

[213], X = D, $n = 1$; [214], X = E, $n = 4$; [215], X = F, $n = 0$; [216], X = F, $n = 1$

[217], X = D; [218], X = E; [219], X = F

A = $CH_2CH_2CH_2$ D = 2,5-furyl

B = CH_2OCH_2 E = 2,5-tetrahydrofuryl

C = 1,3-phenyl F = 2,6-pyridyl

36-membered rings. (*iii*) If one CH_2OCH_2 unit in 18-crown-6 is replaced by A–F, the complex stabilities increase in the order:

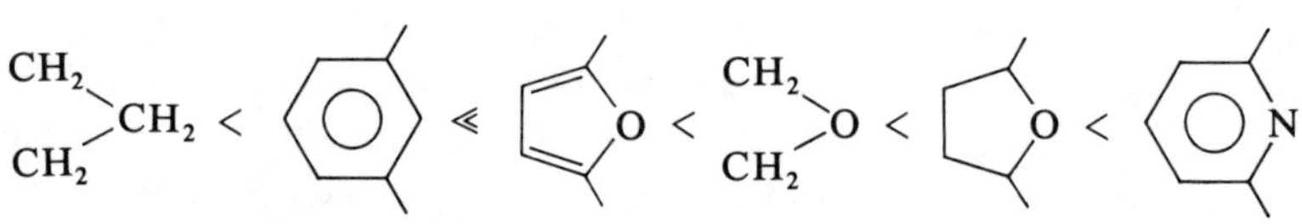

(A) < (C) ≪ (D) < (B) < (E) < (F)

(*iv*) If successively one, two and three furan units are substituted for CH_2OCH_2 in 18-crown-6, the stability sequence is [208] > [211] > [217] > [213]. The difference between the results obtained with the positional isomers [211] and [213] was attributed to the fact that [213] always has one furan group involved in H-bonding, whereas [211] has one out of three rotamers in which only aliphatic oxygen donor sites are used for H-bonding. A similar increased stabilization due to complementary arrangement of binding sites was suggested for the pyridine unit. (*v*) Complexes in which SCN^- is the anion are about 2 kcal mol^{-1} more stable than complexes with Cl^- as the anion.

The same authors also studied the effect of replacing a CH_2CH_2 unit in

TABLE 50

Log *K* values for t-$BuNH_3X$ complexes in $CDCl_3$ at 24 °C[a]

Compound	Ring size	X	Log *K*	Compound	Ring size	X	Log *K*
Pentaglyme [43]	0	SCN	2.11	[212]	18	SCN	6.14
[3]	18	SCN	6.48	[213]	18	SCN	2.41
		Cl	5.05	[214]	36	SCN	4.41
[9]	18	SCN	5.79	[215]	12	SCN	2.89
		Cl	4.28	[216]	18	SCN	6.26
[11]	18	SCN	4.93			Cl	4.67
([20] + [21])	18	SCN	5.56	[217]	18	SCN	3.01
[28]	18	SCN	3.52	[218]	18	SCN	6.62
[79]	18	SCN	6.76	[219]	18	Cl	4.98
		Cl	5.09	[220]	20	SCN	<2
[207]	18	SCN	3.22	[221]	23	SCN	2.11
[208]	18	SCN	5.32	[222]	20	SCN	3.13
		Cl	3.98	[223]	28	SCN	<2
[209]	15	SCN	3.51	[224]	34	SCN	2.41
[210]	18	SCN	6.62	[225]	40	SCN	2.21
[211]	18	SCN	4.14	[226]	22	SCN	<2.11

[a] Data from Timko *et al.*, 1974; Newcomb *et al.*, 1977b; Timko *et al.*, 1977; Helgeson *et al.*, 1977

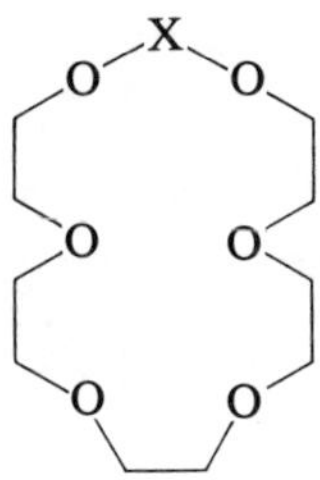

[9], X = G

[220], X = H

[221], X = J

[222], X = K

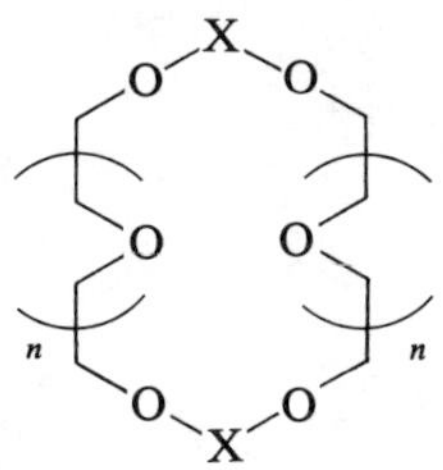

[11], X = G, *n* = 1

[223], X = H, *n* = 2

[224], X = H, *n* = 3

[225], X = H, *n* = 4

([20] + [21]), X = I, *n* = 1

[226], X = K, *n* = 1

G = 1,2-benzo

H = 1,4-benzo

I = 1,2-cyclohexyl

J = 4,12-[2.2]paracyclophanyl

K = 2,2′-(1,1′-binaphthyl)

18-crown-6 by the groups G–K. The results (Table 50) show the following sequence of complex stabilities:

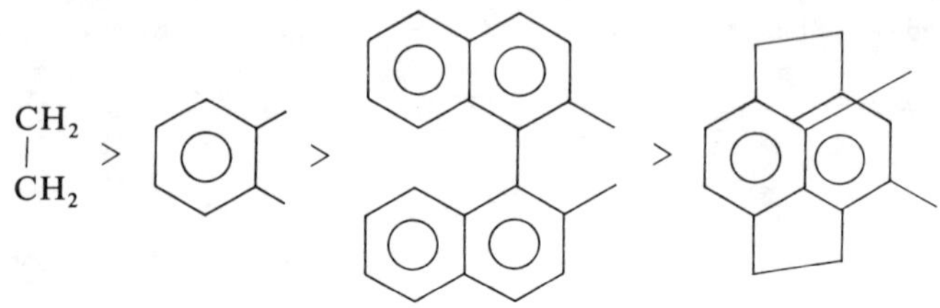

(G) (K) (J)

This order reflects not only the differences in basicity of the binding sites, but also the differences in ring size and ring conformation. [2.2]Paracyclophanes containing divergent crown ether rings (such as [58] and [59]) showed no synergistic effects in binding t-butylammonium picrate (Helgeson *et al.*, 1977). Their complexes with hexamethylenediammonium dihexafluorophosphate are probably polymeric rather than discrete 1 : 1 complexes. However, crown ether [227] containing convergent crown-ether rings did form 1 : 1 complexes with

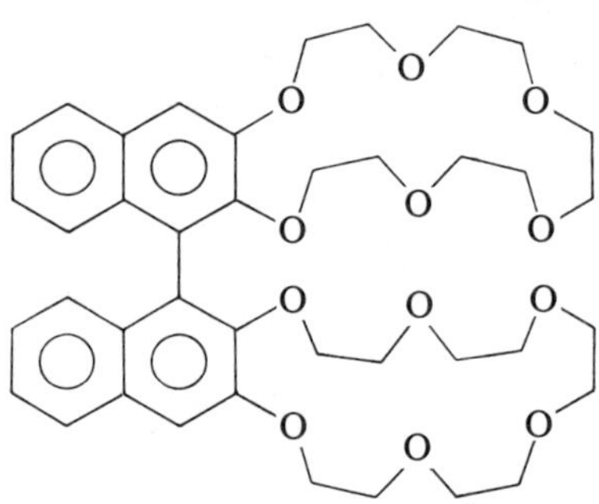

[227]

hexafluorophosphate salts of tetramethylenediammonium or *p*-phenylenediammonium salts (Tarnowski and Cram, 1976).

Coxon *et al.* (1978) have studied the effect of conformational differences on the stability of t-butylammonium thiocyanate complexes in deuterochloroform. The complexing ability of 18-crown-6 [3] towards t-$BuNH_3^+$ is reduced upon fusion with either one or two six-membered rings (Table 51). The authors conclude that any deviation from the flexible all-gauche OCH_2CH_2O conformation results in a loss of binding energy which is enthalpic rather than entropic in origin. For dicyclohexyl-18-crown-6 the observed stability sequence is *cis-anti-cis* > *cis-syn-cis* > *trans-syn-trans*, an order which was correlated with a detailed stereochemical classification. Fusion of glucose, galactose, mannose, and altrose residues to 18-crown-6 (to form compounds [228],

TABLE 51

Log K values for t-$BuNH_3SCN$ complexes in $CDCl_3$ at 25 °C

Crown ether		Log K	Crown ether		Log K
18-Crown-6 [3]		6.48[a]	2,3 : 2′,3′-α,α-D,D-*bis*gluco-18-crown-6	[239]	<1.7[e]
Dicyclohexyl-18-crown-6			2,3 : 3′,2′-α,α-D,D-*bis*gluco-18-crown-6	[240]	<1.7[e]
cis-syn-cis [20]		4.23[b]	2,3 : 2′,3′-α,α-D,D-*bis*galacto-18-crown-6	[241]	3.67[e]
cis-anti-cis [21]		5.95[b]	2,3 : 3′,2′-α,α-D,D-*bis*galacto-18-crown-6	[242]	3.76[e]
trans-syn-trans [22]		3.85[b]		[243]	<1.7[f]
α-D-Gluco-18-crown-6	[228]	3.30[b]		[244]	2.72[f]
	[229]	2.76[g]		[245]	<1.7[h]
β-D-Gluco-18-crown-6	[230]	3.11[b]		[246]	≈3.77[h]
	[231]	3.00[c]		[247]	≈3.16[h]
α-D-Galacto-18-crown-6	[232]	5.30[b]		[248]	5.51[i]
	[233]	4.00[g]		[249]	4.72[i]
β-D-Galacto-18-crown-6	[234]	3.76[b]		[250]	4.23[i]
	[235]	3.62[c]		[251]	4.34[i]
α-D-Manno-18-crown-6	[236]	4.59[b]		[252]	≈5.1[j]
	[237]	3.11[d]		[253]	<4.8[j]
α-D-Altro-18-crown-6	[238]	<1.7[d]			

[a] Timko *et al.*, 1977
[b] Coxon *et al.*, 1978
[c] Pettman and Stoddart, 1979a
[d] Pettman and Stoddart, 1979b
[e] Laidler *et al.*, 1979
[f] Laidler and Stoddart, 1979
[g] Laidler and Stoddart, 1977a
[h] Laidler and Stoddart, 1976; reported values + 0.8 to place them on the same scale
[i] Ando *et al.*, 1978
[j] Curtis *et al.*, 1975b; reported values + 0.8

[230], [232], [234], [236], and [238]) also results in a marked drop in binding energy. In general, compounds with *trans* ring junctions form weaker complexes that those with *cis* ring junctions. The exceptional stability of the galacto-compounds [232] compared to that of gluco-crown ethers [228] was attributed to participation of ether oxygens of the galacto residue in hydrogen

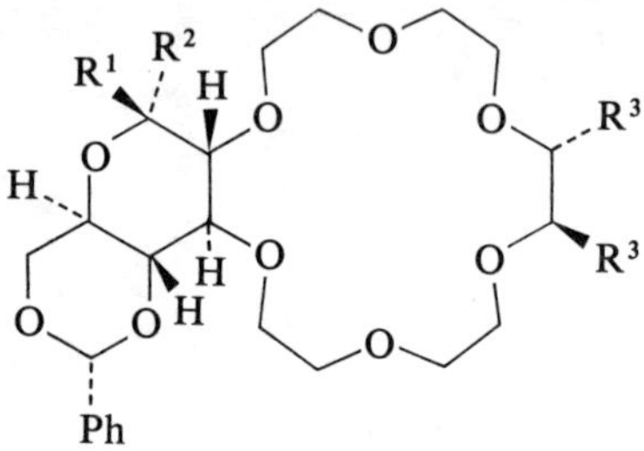

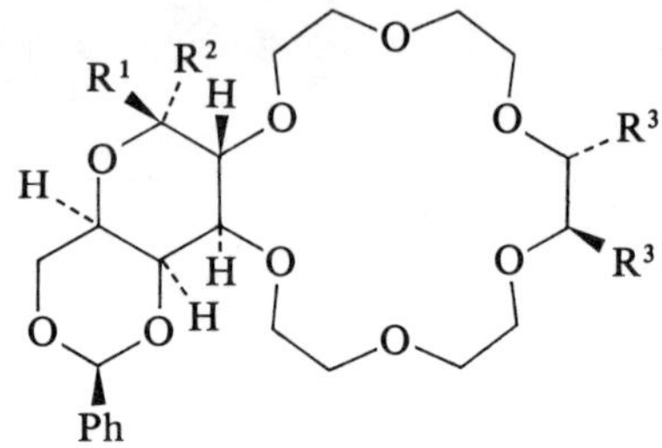

	R^1	R^2	R^3	
[228]	H	OMe	H	[232]
[229]	H	OMe	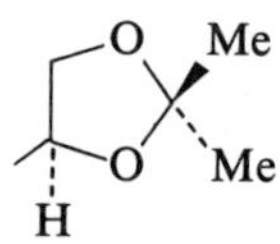	[233]
[230]	OMe	H	H	[234]
[231]	OMe	H	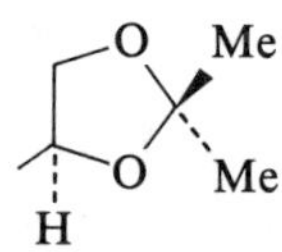	[235]

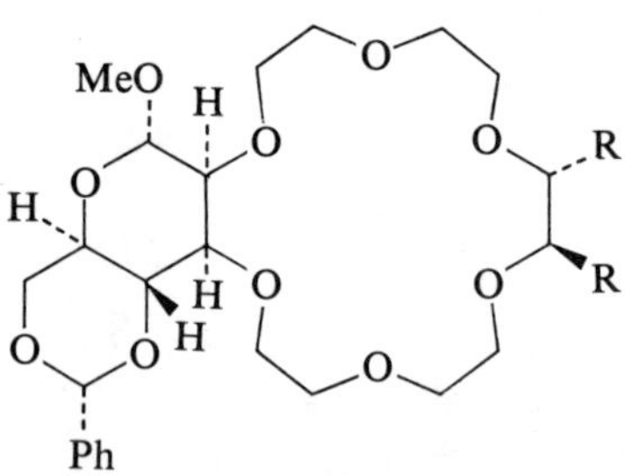

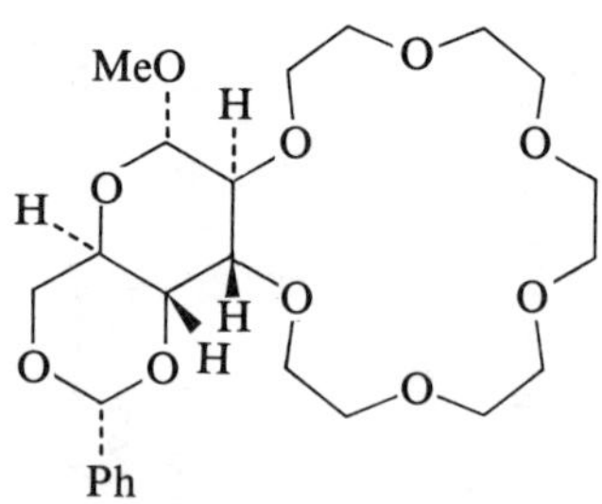

[236], R = H

[237], R = 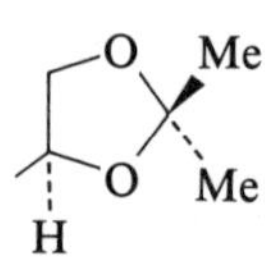

[238]

bonding and/or electrostatic interaction with the alkylammonium cation (Laidler and Stoddart, 1977a). The introduction of 2,2-dimethyl-1,3-dioxolanyl groups to form compounds [229], [231], [233], [235], and [237] decreases the stability of the complex. This observation was attributed to increased steric interactions in the complex. The same explanation was given for the low stability of t-$BuNH_3SCN$ complexes with *bis*gluco- and *bis*galacto-18-crown-6 derivatives [239]–[242] (Laidler *et al.*, 1979), and the weak complexing ability of the 22-crown-6 [243] and the 20-crown-6 derivatives [244] (Laidler and Stoddart, 1979).

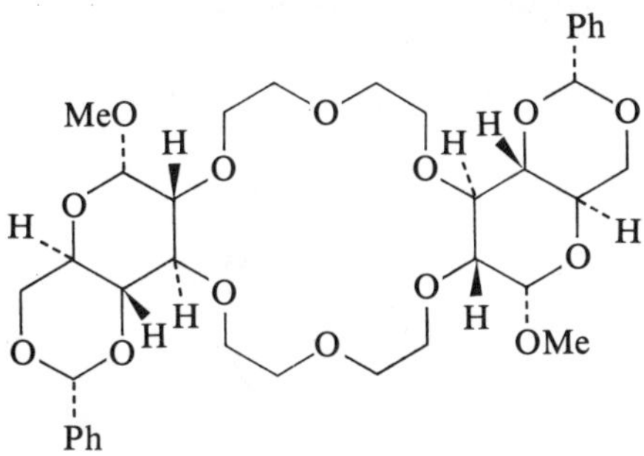

[239]

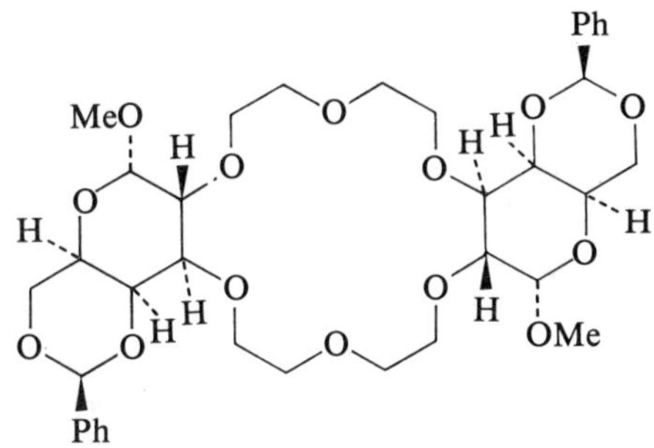

[241]

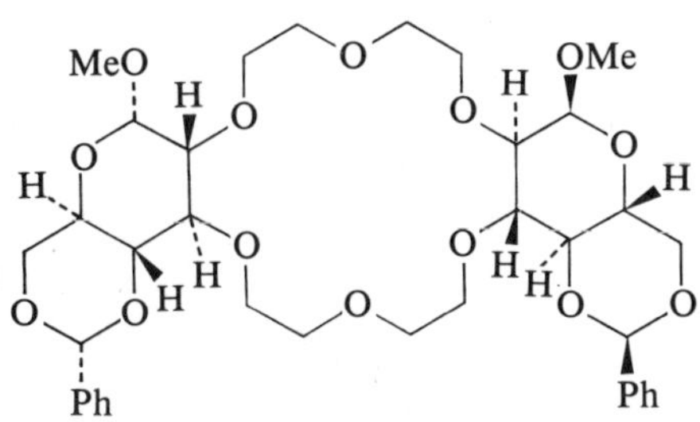

[240]

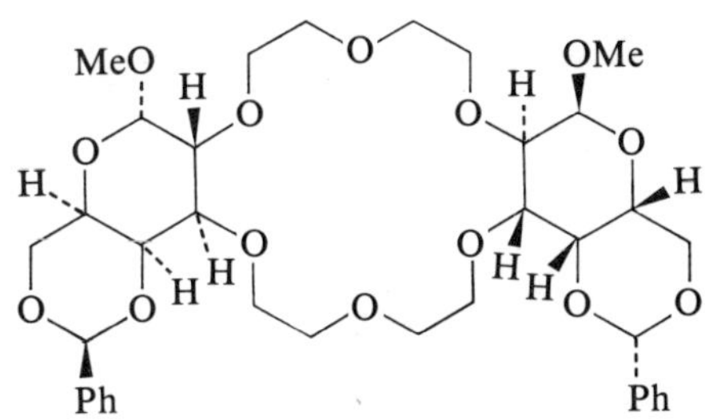

[242]

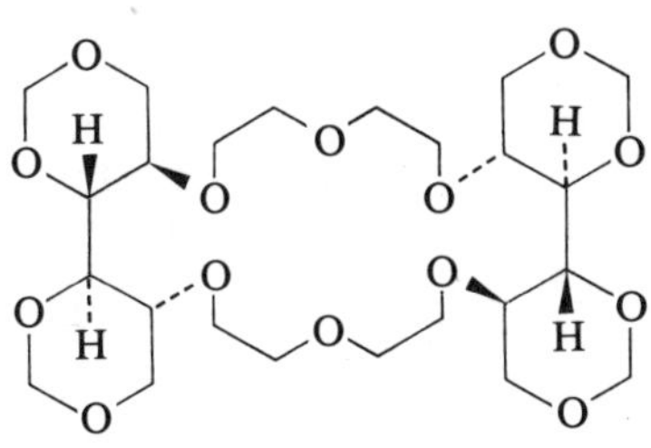

[243]

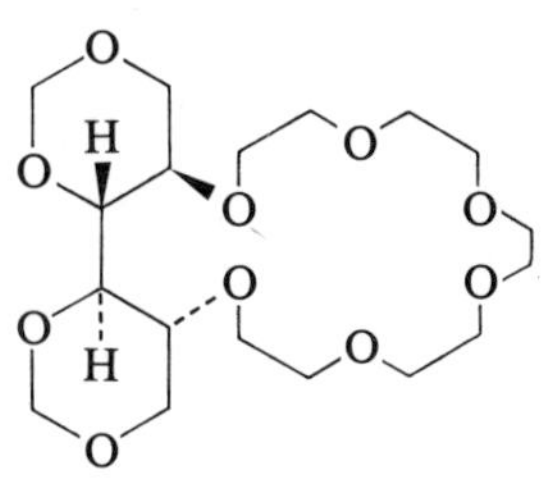

[244]

The crown ethers based on carbohydrates differ from "simple" crown ethers in two important respects. Firstly, all of these crown ethers except [240], [242],

[243], and [244] lack a C_2-axis of symmetry and therefore possess two different faces with different complexing abilities and selectivities towards alkylammonium cations. From ^{1}H-nmr studies the relative population of both faces can be determined (Pettman and Stoddart, 1979a, 1979b). Secondly, many of the crown ethers [228]–[244] form much more stable complexes with benzylammonium than t-butylammonium salts; the difference may be as large as 6 kcal mol^{-1}. A stabilizing dipole-induced dipole interaction between the phenyl group in the cation and the 1,3-dioxan rings in the crown ether was proposed to explain this difference (Laidler and Stoddart, 1977a).

The introduction of four 2,2-dimethyl-1,3-dioxolanyl groups into 18-crown-6 to form compound [245] causes the log K value to drop from 6.48 to below 1.7 (Laidler and Stoddart, 1976). Replacement of two CH_2OCH_2 units

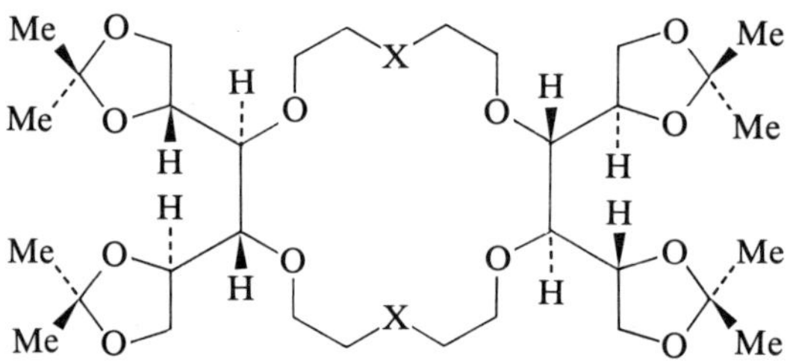

[245], X = O

[246], X = NCH_3

[247], X = 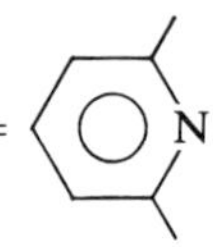

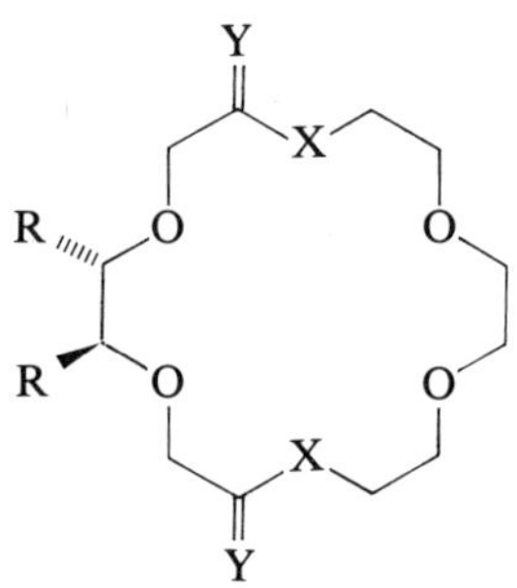

	X	Y	R
[248]	O	H_2	$PhCH_2OCH_2$
[249]	O	H_2	$CH_3\underset{\underset{O}{\|}}{C}OCH_2$
[250]	NH	O	$PhCH_2OCH_2$
[251]	NH	H_2	$PhCH_2OCH_2$

by pyridyl units, or substitution of $N(CH_3)$ for O increases the complexing ability, although the complexes of [247] are still much weaker than those of the unsubstituted *bis*-pyridyl-18-crown-6 [216]. The effect of replacing an ether–oxygen binding site by an amine–nitrogen site is wholly dependent on the total structure of the crown ether, however, as is evident from the work of Ando *et al.* (1978), who showed the all-oxygen crown ether [248] to form more stable complexes than the corresponding diaza-compound [251]. In these series, there is little difference in complexing ability between diamide [250] and diamine [251]. Incorporation of four benzyloxymethyl and trityloxymethyl groups into 18-crown-6 (giving compounds [252] and [253] respectively)

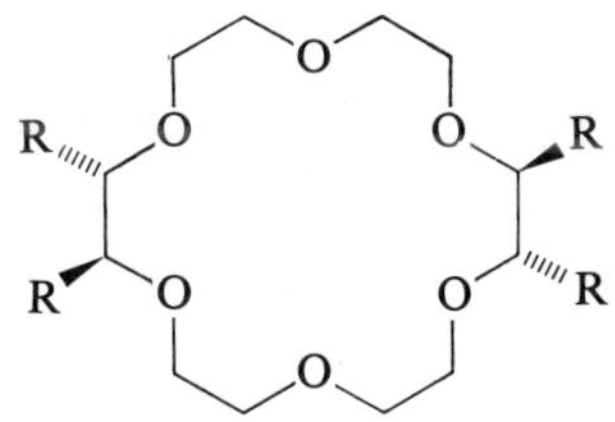

[252], R = CH_2OCH_2Ph [254], X = $CON(CH_3)_2$

[253], R = $CH_2OC(Ph)_3$ [255], X = CO_2^-

decreases the complexing ability towards t-BuNH_3^+ as a result of increased steric repulsion (Curtis *et al.*, 1975b).

Behr *et al.* (1976) reported that with NH_4^+ cations in water simple 18-crown-6 derivatives such as [254] form only very weak complexes (log $K < 0.7$). A large increase in complex stability was observed on introducing carboxylate groups close to the cavity. Values of log K for compound [255] and NH_4^+ and $CH_3NH_3^+$ were 3.5 and 2.9 respectively. The extra binding energy was attributed to electrostatic and hydrophobic effects. In apolar solvents, anionic binding sites also show stabilizing effects. McKervey and Mulholland (1977) observed that 1,3-xylyl-18-crown-5 bearing an intra-annular OH group ([256], $n = 3$, R_1 = H, R_2 = OH) formed stable 1:1 complexes with ammonia. Newcomb *et al.* (1977a) reported stable 1:1 complexes between t-butylamine and 1,3-xylyl-18-crown-5 bearing an intra-

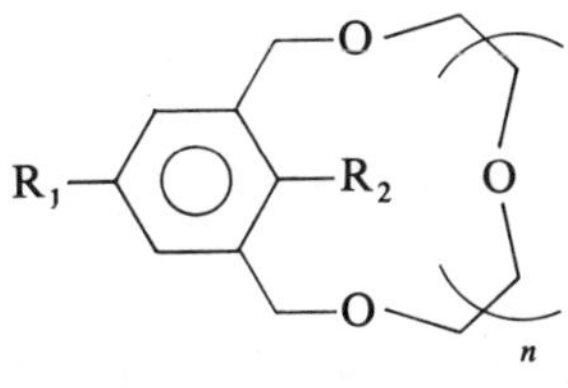

[256]

annular carboxylate group ([256], $n = 3$, R_1 = H, R_2 = CO_2H). Many intra-annular substituents decrease the complex stability as compared with the parent compound (Table 52). Apparently the introduction of an extra binding site does not outweigh the increased steric shielding of the cavity, except for the CO_2CH_3 group which gives a 16-fold increase in binding constant.

The effect of remote substitutents on the complexing ability of crown ethers [256] towards t-butylammonium salts has been studied by Moore *et al.* (1977). The results (Table 52) show that electron-attracting substituents decrease the stability of the complex. The t-BuNH_3^+ cation is even more sensitive to remote substituent effects than alkali-metal cations. This fact was attributed to the non-polar character of the alkyl group and to the different structure of the

TABLE 52

Effect of remote and intra-annular substituents on log K values for t-$BuNH_3X$ complexes of 1,3-xylyl-crown ethers [256] in $CDCl_3$ at 24°C[a]

			X		
n	R_1	R_2	SCN^-	ClO_4^-	picrate
2	H	CO_2CH_3	2.69	—	—
3	H	CO_2CH_3	4.75	—	—
4	H	CO_2CH_3	3.48	—	—
7	H	CO_2CH_3	2.89	—	—
3	H	CO_2H	3.09	—	—
3	H	CH_2OH	3.36	—	—
3	H	CH_2OCH_3	3.40	—	—
3	H	H	3.52	5.3	3.5
3	H	CN	3.06	—	—
3	Br	H	2.5	4.4	2.9
3	t-Bu	H	3.7	5.6	3.7
3	$CO_2C_2H_5$	H	2.8	—	—
3	OCH_3	H	3.5	5.4	—
3	SCH_3	H	—	4.8	—
3	CN	H	—	3.8	—

[a] Data taken from Newcomb *et al.*, 1977a; Moore *et al.*, 1977

complex. From the data in Table 52 it also follows that the complex stability decreases as a function of the anion in the order $ClO_4^- > SCN^- \approx$ picrate.

The determination of binding constants from extraction experiments provides a rapid and easy way of obtaining information about the complexing ability of crown ethers. Strictly speaking, however, the binding constants (both absolute and relative) are meaningful only in situations in which an excess of water is available. Just like built-in additional binding sites, water molecules can affect the stability of a complex by acting either as a H-bond donor or as a H-bond acceptor towards anion, cation or crown ether. As a result, the (relative) complexing abilities of crown ethers may change considerably on going from two-phase to anhydrous conditions. This was illustrated by De Jong *et al.* (1976a, 1976b) for a series of 1,3-xylyl-crown ethers ([256], $R_1 = R_2 = H$, $n = 2$–7). Partition experiments with t-$BuNH_3PF_6$ using low concentrations of crown ether in the chloroform phase gave a single sharp maximum in binding capacity for the 18-membered ($n = 3$) crown ether (Fig. 3). On increasing the crown concentration, a second maximum was observed for the 24-membered compound. Furthermore, the 18-membered compound did not co-extract water when binding t-$BuNH_3^+$, whereas with the 24-

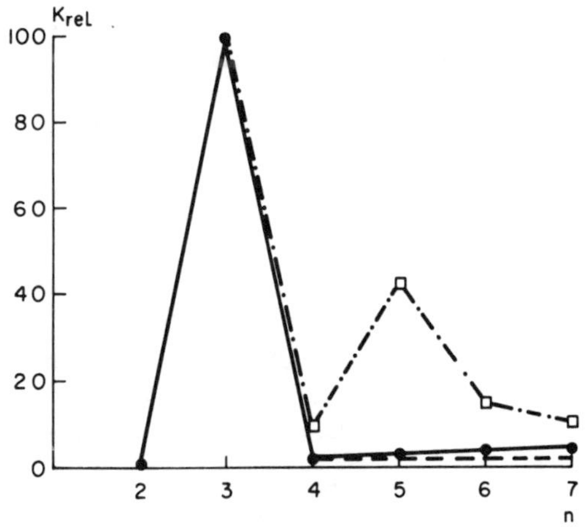

FIG. 3 Comparison of relative K_a-values obtained under anhydrous conditions (——), and from partition experiments at low (---, 0.01 M) and high (—·—, 0.5 M) crown-ether concentration. Salt, t-$BuNH_3PF_6$; crown ethers, [256], $R_1 = R_2 = H$. Data from De Jong *et al.* (1976a, 1967b)

membered compound each t-$BuNH_3^+$ cation extracted was accompanied by about two water molecules. For the same series of crown ethers, the relative binding constants in anhydrous chloroform were determined by a ^{1}H-nmr titration technique (De Jong *et al.*, 1976b). Under these conditions, a single maximum for the 18-membered ring was observed, even at high crown-ether concentrations. These observations were rationalized by proposing a scheme according to which crown ethers having too large a cavity to accommodate t-$BuNH_3^+$ form hydrated dimers in the presence of water.

There is at present no general method available to determine association constants of crown ethers and protonated amines in anhydrous apolar solutions. The UV-spectroscopic method used by Behr and Lehn (1978) can only be applied to complexes in which charge-transfer interactions occur between π-systems in the cation and crown ether. A more general but also more laborious method consists in determining the binding constants from the rates of cation exchange (De Jong *et al.*, 1976c), a method which is described in the next section.

KINETIC STABILITY

De Jong *et al.* (1976c, 1977) have studied the rate of decomplexation of t-$BuNH_3PF_6$ (AM) complexes of crown ethers. The exchange broadening of the t-butyl nmr signal was used to determine the rate of cation exchange between a kinetically stable (X.AM) and a kinetically unstable (Y.AM)

$$X.AM \underset{k_1}{\overset{k_1}{\rightleftarrows}} X + AM \tag{45}$$

$$Y.AM \underset{k_2}{\overset{k_2}{\rightleftarrows}} Y + AM \tag{46}$$

$$X.AM + AM^* \xrightarrow{k_3} X.AM^* + AM \tag{47}$$

complex. The results were consistent with a mechanism in which cation exchange occurs by (45)–(47), the reciprocal lifetime of X.AM being given by (48). Since $[AM] = \{[Y.AM]/K_2\}^{1/2}$ for solutions containing the complexes X.AM and Y.AM only, and $[AM] = [Y.AM]/K_2.[Y]_e$ for solutions containing an excess of free crown ether Y ($[Y]_e$), equation (48) yields both k_1

$$\tau^{-1}_{X.AM} = k_1 + k_3[AM] \tag{48}$$

and K_2 by extrapolation to $[AM] = 0$. The results (Table 53) show that: (*i*) the relative values of log K for compounds [3], [9], [11], [28] and [208] as determined from cation exchange experiments follow the same order as the log K values determined by extraction (Table 50); (*ii*) the rates of decomplexation are very sensitive to changes in crown ether structure, whereas the rates of complexation are almost constant [$(0.8–1.5) \times 10^9$ M^{-1} s^{-1}] and probably diffusion controlled. Consequently, the observed differences in the thermodynamic stability of the complexes are fully reflected in the different rates of decomplexation.

Another method for determining the rates of decomplexation consists in

TABLE 53

Kinetic and thermodynamic data for $t\text{-}BuNH_3PF_6$ complexes of crown ethers in $CDCl_3$ at 20°C[a]

Crown ether	k_1 /s^{-1}	$\Delta G^{\ddagger}_{d}$ /kcal mol^{-1}	$10^{-9}k_{-1}$ /M^{-1} s^{-1}	Log K	$-\Delta G^0$ /kcal mol^{-1}
18-Crown-6 [3]	65	14.7	1.5	7.36	9.8
Benzo-15-crown-5 [8]	—	—	—	3.91	5.2
Benzo-18-crown-6 [9]	155	14.2	0.8	6.70	8.9
Dibenzo-18-crown-6 [11]	850	13.2	1.0	6.08	8.1
1,3-Xylyl-18-crown-5 [28]	5400	12.1	1.2	5.34	7.1
1,3-Xylyl-21-crown-6 [256; $R_1 = R_2 = H$, $n = 4$]	—	—	—	4.46	5.9
1,2-Xylyl-17-crown-5 [27]	9000	11.8	1.3	5.15	6.9
2,5-Furyl-18-crown-6 [208]	1100	13.0	1.4	6.11	8.2
3,4-Furyl-17-crown-5 [33]	7000	12.0	0.8	5.04	6.7

[a] De Jong *et al.*, 1976c; 1977

measuring the rates of cation exchange between the opposite faces of the crown ether (49) at the coalescence temperature for signals H_a and H_b. It was shown

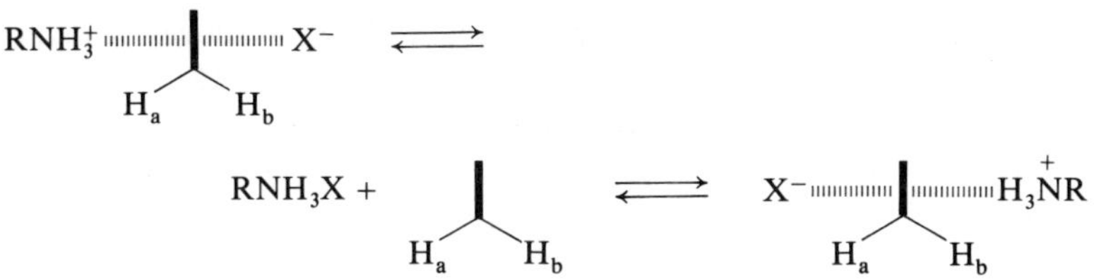

(49)

for 1,3-xylyl-18-crown-5 [28], 1,2-xylyl-17-crown-5 [27], 2,5-furo-18-crown-6 [208], and 3,4-furo-17-crown-5 [33] that within experimental error this method yields the same rates of decomplexation as cation exchange between different crown ethers (De Jong *et al.*, 1977). 1,3-Xylyl-18-crown-5 [28] is particularly suitable for studying the cation exchange between the front and back faces, since the free host is virtually strain-free in a planar conformation, and therefore the measured free energy of activation for decomplexation, $\Delta G_d^\ddagger$, includes few if any ring inversion components. The effect of the organic group R on the kinetic stability of RNH_3PF_6 complexes of 1,3-xylyl-18-crown-5 is shown in Table 54. It is seen that the free energies of activation for the

TABLE 54

Effect of cation structure on the kinetic stability of RNH_3PF_6 complexes of [28] in $CDCl_3$[a,b]

R	T_c /°C	$\Delta G_d^\ddagger$ /kcal mol^{-1}	R	T_c /°C	$\Delta G_d^\ddagger$ /kcal mol^{-1}
CH_3	−43	11.4	Cyclopropyl	−22	12.5
CH_3CH_2	−33	11.9	Cyclopentyl	−19	12.7
$(CH_3)_2CH$	−25	12.3	$C_6H_5CH_2CH_2$	−31	12.3
$(CH_3)_3C$	−27	12.3	$C_6H_5CH_2$	−16	12.9
$(CH_3)_3CCH_2$	−25	12.3	*p*-$CH_3C_6H_4$	−7	13.1
$(CH_3CH_2)_2CH$	−36	11.8	*p*-$CH_3OC_6H_4$	−5	13.1

[a] $\Delta G_d^\ddagger$ values given at the coalescence temperature T_c
[b] De Jong *et al.*, 1979

decomplexation step increase in the order $CH_3NH_3^+ < CH_3CH_2NH_3^+ < (CH_3)_2CHNH_3^+$, while the isopropyl and t-butylammonium complexes are kinetically equally stable. Hence, α-methyl substituents induce hardly any steric repulsion in the complex. This also follows from the fact that the cyclopropylammonium complex is only slightly more stable than the isopropylammonium complex. Steric hindrance of β-substituents does not seem to be significant either. Only for $(CH_3CH_2)_2CHNH_3^+$ does the effect of steric repulsion become visible, since both isopropyl and cyclopentylammonium are kinetically more stable. The benzylammonium complex was found to be the most stable alkylammonium complex, being only slightly less stable than anilinium complexes. It is interesting to note that 18-crown-6 showed a higher selectivity for alkylammonium ions in methanol solution (Table 49).

The effect of the anion on the kinetic stability of RNH_3X complexes of [28] has also been determined (De Jong *et al.*, 1979). The structure of the cation has little effect on the relative complex stabilities with different anions. For the benzylammonium cation, the following $\Delta G_d^{\ddagger}$-values were observed: PF_6^-, 12.9 (−16°C); ClO_4^-, 11.9 (−35°C); BF_4^-, 11.7 (−39°C); I^-, 11.2 (−48°C); SCN^-, 10.8 (−56°C); Cl^-, <10.2 kcal mol^{-1} (<−70°C). Assuming that for this cation too the rates of complex formation are diffusion controlled, it follows that the thermodynamic complex stabilities decrease in the order $PF_6^- > ClO_4^- > BF_4^- > I^- > SCN^- > Cl^-$. This order is just the reverse of the order of the hydrogen-bonding abilities of these anions in apolar solvents (Arnett *et al.*, 1977). Since the hydrogen bonding between anion and protonated amine must at least partially be disrupted for complexation, it is obvious that the most stable and best structured complexes are formed with anions that are essentially non-hydrogen bonding.

Stoddart and co-workers have reported the free energies of activation for the decomplexation of alkylammonium complexes of crown ethers based on carbohydrates. Since most compounds possess unequal faces, two different complexes can be formed. The $\Delta G_d^{\ddagger}$-values in Table 55 refer to the decomplexation of the thermodynamically more stable complex. The main conclusions from their work can be summarized as follows: (*i*) The structure of the cation has a marked effect on the free energies of decomplexation, $\Delta G_d^{\ddagger}$. For compound [244], $\Delta G_d^{\ddagger}$ decreases in the order $PhCH_2NH_3^+ >$ i-$PrNH_3^+ >$ t-$BuNH_3^+ > CH_3NH_3^+$ when ClO_4^- is the anion. (*ii*) The sequence of cation stabilities is dependent on the type of anion. For SCN^- complexes of [244] $\Delta G_d^{\ddagger}$ decreases in the order: i-$PrNH_3^+ > PhCH_2NH_3^+ \gg$ t-$BuNH_3^+$, $CH_3NH_3^+$. (*iii*) In most cases perchlorates are kinetically more stable than thiocyanates by 0.8–1.8 kcal mol^{-1}. This is not a general rule, however, since the β-D-galacto compound [230] forms more stable complexes with (*S*)-$PhCH(CH_3)$-NH_3SCN than with the corresponding perchlorate. (*iv*) From a comparison of the $\Delta G_d^{\ddagger}$-values (Table 55) with the free energies of complex formation (ΔG_0, Table 51) it follows that the free energies of association ($\Delta G_a^{\ddagger}$) are low (2–4 kcal mol^{-1}) for compounds whose cavities are not sterically shielded (such as [228], [232], [233], [244], [247]). When stereochemical requirements become more stringent (as in the *bis*-gluco- and *bis*-galacto-18-crown-6 derivatives [239]–[242], compounds [237], [238], and [229]), $\Delta G_a^{\ddagger}$ increases to values up to 7.5 kcal mol^{-1} (for [229]). The authors conclude that steric requirements can lower the rates of association considerably, and that bond-making rather than

TABLE 55

Free energies of activation ($\Delta G_d^\ddagger$ in kcal mol^{-1}) for decomplexation of RNH_3X complexes in CD_2Cl_2

Compound	X	R: (*S*)-$PhCHCH_3$	(*R*)-$PhCHCH_3$	$PhCH_2$	t-Bu	Ref.
[228]	SCN^-	11.8	—	11.4	—	*a*
[229]	SCN^-	12.1	11.8	12.1	—	*a*
[230]	SCN^-	11.7	10.5	12.0	10.0	*b*
	ClO_4^-	10.6	—	—	—	*b*
[232]	SCN^-	12.1	—	11.9	10.3	*a*
[233]	SCN^-	11.7	—	12.0	—	*a*
[234]	SCN^-	9.7	10.3	10.2	—	*b*
	ClO_4^-	10.2	—	10.6	—	*b*
[235]	SCN^-	10.6	—	10.8	—	*b*
[236]	SCN^-	11.2	11.2	11.1	—	*c*
	ClO_4^-	—	—	13.0	—	*c*
[237]	SCN^-	12.2	11.5	14.0	10.3	*c*
	ClO_4^-	—	—	14.0	—	*c*
[238]	SCN^-	—	—	—	8.3	*c*
[239]	ClO_4^-	9.9	11.1	10.4	—	*d*
[240]	ClO_4^-	9.2 (−92)	9.4 (−80)	9.3 (−88)	—	*d*
[241]	ClO_4^-	12.5	9.9	12.4	—	*d*
[242]	ClO_4^-	10.0 (−78)	10.0 (−73)	11.9 (−36)	—	*d*
[244]	SCN^-	<8.8	<8.8	8.8 (−90)	<8.2	*e*
	ClO_4^-	10.1 (−65)	—	10.5 (−65)	9.2 (−91)	*e*
[247]	SCN^-	—	—	12.2 (−35)	—	*f*

[a] Laidler and Stoddart, 1977a
[b] Pettman and Stoddart, 1979a
[c] Pettman and Stoddart, 1979b
[d] Laidler *et al.*, 1979
[e] Laidler and Stoddart, 1979
[f] Laidler and Stoddart, 1976

diffusion becomes the rate-limiting step. (*v*) Some of the 18-crown-6 derivatives show modest chiral recognition in the transition state for decomplexation, as is apparent from the $\Delta G_d^\ddagger$ values for (*R*)- and (*S*)-α-phenylethylammonium salts.

For crown ethers in which the ^{1}H-nmr probes H_a and H_b are diastereotopic in the free crown ether too, the cation exchange process (50) becomes more complicated. For 1:1 complexes the measured free energy of activation also involves conformational inversion components ($\Delta G^\ddagger = \Delta G_d^\ddagger + \Delta G_i^\ddagger$), whereas exchange between free and complexed crown ether (2:1 ratio of crown ether to salt) only involves the free energy of activation for decomplexation ($\Delta G_d^\ddagger$). Bradshaw *et al.* (1979b) have recently reported $\Delta G^\ddagger$-values for cation exchange in RNH_3X complexes of crown ethers [257]. The $\Delta G_i^\ddagger$ term was estimated to be 0.7, 0.5, and 1.2 kcal mol^{-1} for compounds [257; $n = 1$] with

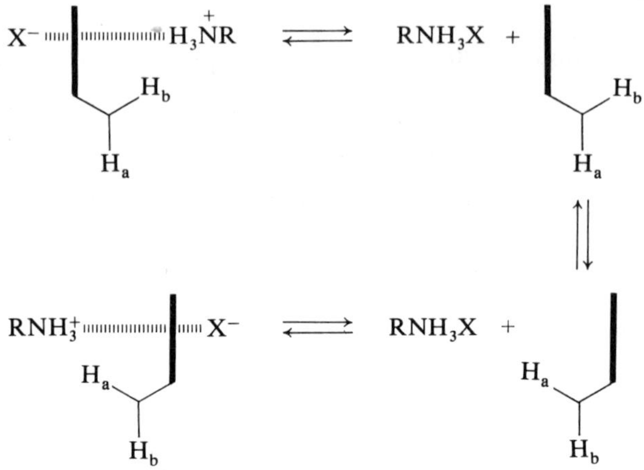

(50)

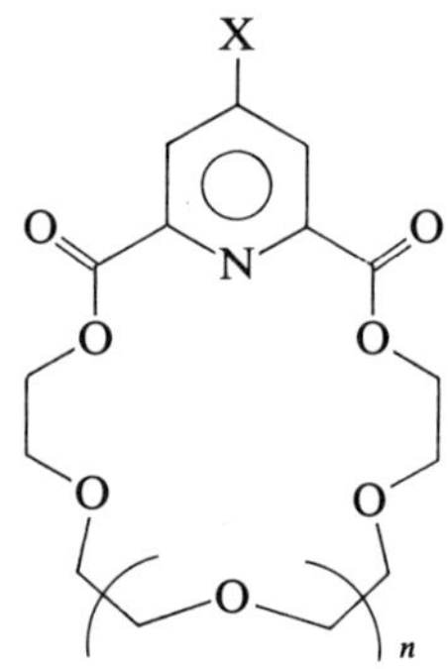

[257]

X = H, Cl, and OCH_3, respectively. Hence, for the 18-membered rings the data in Table 56 indicate that: (*i*) the order of kinetic stability is $OCH_3 > H >$ Cl, which parallels the order of basicity of 4-substituted pyridines; (*ii*) benzylammonium forms more stable complexes than t-butyl- and isopropylammonium, but the relative stability of the latter depends on the substituent; (*iii*) the complexes of the 21-membered compounds are considerably less stable, and show a larger difference between $PhCH_2NH_3^+$ and t-$BuNH_3^+$ cations. The contribution of ring inversion is not known for the 21-membered rings, however. The difference in kinetic stability between $PhCH_2NH_3ClO_4$ and $PhCH_2NH_3SCN$ complexes of the 18-membered ring compounds was 2.4–2.8 kcal mol^{-1} in CD_2Cl_2, which is considerably larger than the anion effect in $CDCl_3$ for 1,3-xylyl-18-crown-5 [28].

Sutherland and co-workers have studied the cation-exchange processes in alkylammonium complexes of crown ethers [258]–[262]. Depending on the position of the exchanging ^{1}H-nmr sites, the measured $\Delta G^{\ddagger}$-values may refer to three different types of cation exchange: (*i*) cation exchange between free and

TABLE 56

Substituent effect on the kinetic stability ($\Delta G^{\neq}$ in kcal mol^{-1} at T_c) of RNH_3ClO_4 complexes of crown ethers [257] in CD_2Cl_2[a]

		R		
n	X	$PhCH_2$	t-Bu	i-Pr
1	H	13.0 (+10)	11.8 (−20)	12.2 (−10)
	Cl	12.5 (0)	11.3 (−30)	11.7 (−20)
	OCH_3	13.7 (+25)	12.3 (−10)	12.2 (−10)
2	H	11.6 (−30)	9.1 (−85)	—
	Cl	12.0 (−25)	8.9 (−90)	—
	OCH_3	12.6 (−15)	9.5 (−75)	—

[a] Bradshaw *et al.*, 1979b

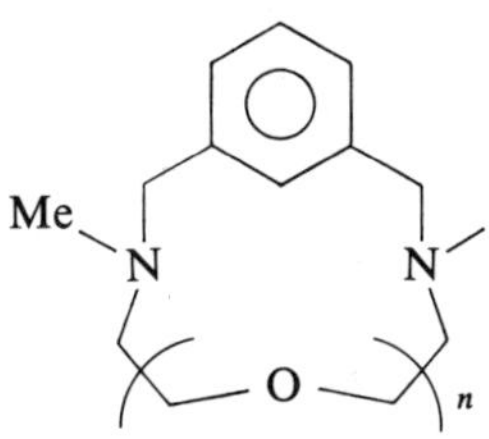

[258]

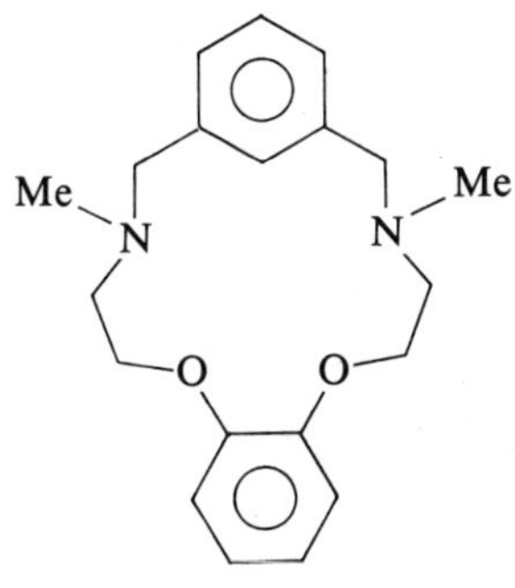

[259]

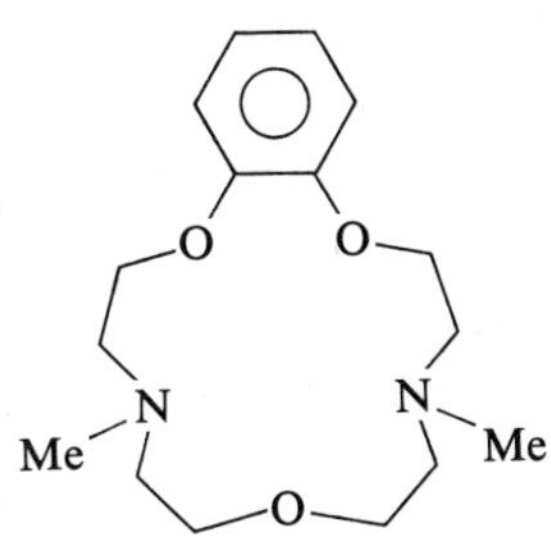

[260]

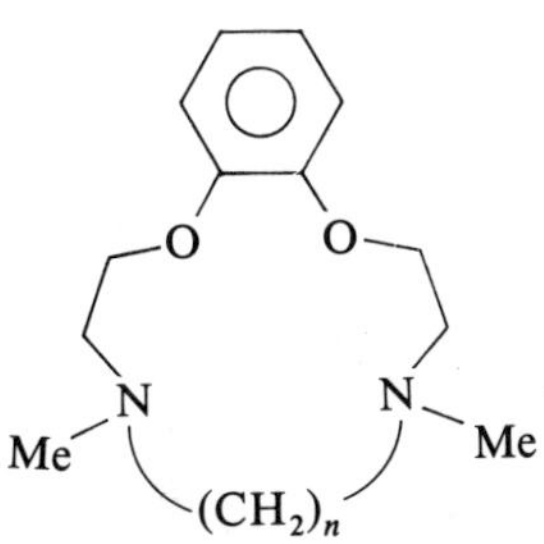

[261]

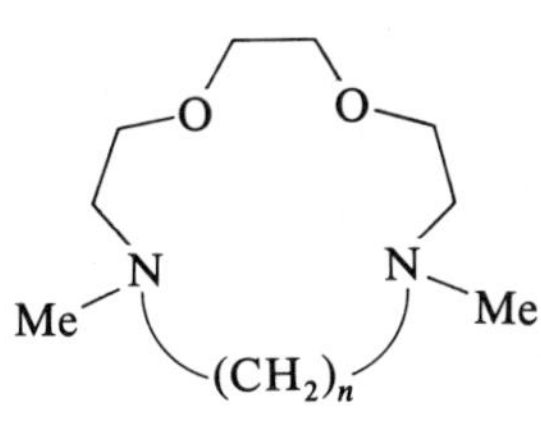

[262]

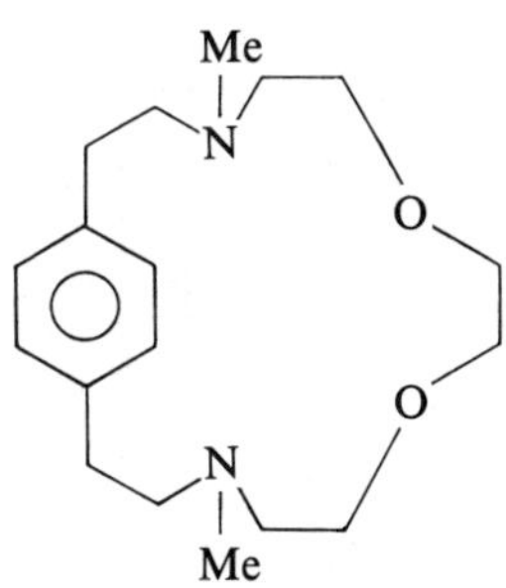

[263]

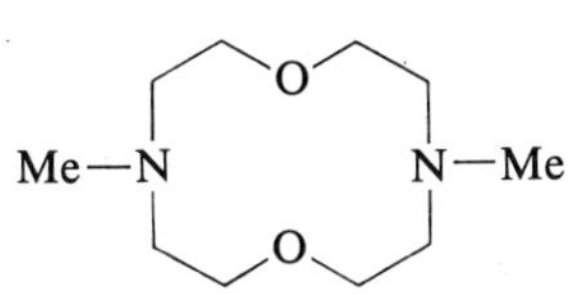

[264]

TABLE 57

Free energies of activation ($\Delta G^{\neq}$ in kcal mol^{-1} at T_c) for exchange of RNH_3X in diaza-crown ether complexes in CD_2Cl_2

Compound	X	R: PhCHCH$_3$	PhCH$_2$	t-Bu	i-Pr	Et	Me	Ref.
[258], $n = 2$	SCN$^-$	(*R*, *S*): 10.9 (−45)	10.8 (−45)	9.3 (−80)	10.1 (−62)	10.7 (−50)	10.1 (−61)	h,i
$n = 3$	SCN$^-$	(*R*, *S*): 9.8 (−65)	9.9 (−65)	—	—	—	9.7 (−61)	h
[259]	SCN$^-$	—	11.1 (−40)	9.1 (−82)	10.2 (−57)	10.0 (−62)	—	i
[260]	SCN$^-$	(*R*): 12.6 (−5)[a]	13.1 (0)	—	—	—	—	i
[261], $n = 4$	SCN$^-$	—	8.8 (−90)	—	—	—	—	j
$n = 5$	SCN$^-$	—	11.4 (−35)	—	—	—	—	j
$n = 6$	SCN$^-$	—	10.4 (−60)	—	—	—	—	j
$n = 7$	SCN$^-$	—	10.1 (−60)	—	—	—	—	j
$n = 8$	SCN$^-$	—	9.5 (−70)	—	—	—	—	j
[262], $n = 4$	SCN$^-$	—	10.4 (−55)	—	—	—	—	j
$n = 5$	SCN$^-$	—	10.7 (−55)	—	—	—	—	j
$n = 6$	SCN$^-$	—	10.0 (−65)	—	—	—	—	j
$n = 7$	SCN$^-$	—	9.4 (−80)	—	—	—	—	j
[263]	SCN$^-$	(*S*): 10.7 (−70)	11.0 (−50)	—	—	—	12.1 (−30)	k
	ClO_4^-	—	11.5 (−50)[d]	10.2 (−65)[c]	11.3 (−50)	—	[b]	k
[264]	SCN$^-$	(*S*): 13.1 (−5)[e]	13.4 (0)	11.7 (−35)	12.8 (−12)	—	11.0 (−50)	l
	ClO_4^-	(*S*): 13.2 (−8)[f]	11.7 (−30)	12.8 (−21)[g]	13.4 (−12)	—	10.2 (−64)	l

$\Delta G_d^{\neq}$ values (T_c) are: [a] 11.4 (−47); [b] 11.3 (−56); [c] 9.3 (−90); [d] 10.7 (−40); [e] 10.2 (−47); [f] 10.3 (−52); [g] 10.1 (−64)

[h] Leigh and Sutherland, 1975

[i] Hodgkinson *et al.*, 1976a

[j] Hodgkinson *et al.*, 1976b

[k] Beckford *et al.*, 1978

[l] Metcalfe *et al.*, 1977

complexed crown ether; (*ii*) cation exchange between the back and face sides of the complex, a process which involves inversion components, and (*iii*) in the case of a racemic guest, an exchange between (*R*)- and (*S*)-cations on the same side of the crown ether. Unless stated otherwise, the data in Table 57 refer to cation exchange accompanied by ring and/or nitrogen inversion. The effect of cation structure on the kinetic stability of complexes of [258; $n = 2$] shows that both α- and β-substituents induce steric repulsion in the complex. In the series R = H, Me, Et, i-Pr, t-Bu, $\Delta G^{\ddagger}$ first increases to reach a maximum for R = Et and then decreases. The introduction of β-methyl substituents invariably leads to decreased kinetic stability (Hodgkinson *et al.*, 1976a). The cavity of compound [259] is apparently less sterically shielded, since the isopropylammonium forms the most stable complex in this case.

The effect of ring size is visible from the data for compounds [261] and [262]. It follows that the most stable complexes are formed by 15- and 16-membered rings. To account for the dependence of $\Delta G^{\ddagger}$ on guest and host structures, the authors proposed a model involving only two hydrogen bonds between the cation and crown ether and one hydrogen bond between the cation and the thiocyanate anion. For the diazaparacyclophane [264], Beckford *et al.* (1978) proposed in addition a stabilizing interaction between the π-electron system of the phenylene ring and the cation.

A two-point binding model was also proposed by Metcalfe *et al.* (1977) to account for the kinetic stability of complexes of the diaza-12-crown-4 derivative [264]. For R_2NH_2X complexes, perchlorates are more stable than thiocyanates, whereas for RNH_3X complexes the thiocyanates are the more stable for R = CH_3, Et, $PhCH_2$ and perchlorates are the more stable when R = i-Pr, t-Bu, and $PhCHCH_3$. Hydrogen bonding between cation and anion will decrease the stability of complexes with secondary alkylammonium salts since both N–H bonds of the guest are needed to bind the host. In the case of primary alkylammonium salts, one N–H bond of the guest is still available for H-bonding with the anion, and therefore thiocyanates may be more stable than perchlorates.

6 Complexes of racemic salts with chiral crown ethers

GENERAL

In the preceding section it was shown that the stability of crown-ether complexes with alkylammonium salts depends on the relationship between the structures of the crown ethers and the ammonium ions. How critically this relationship determines the complex stability will become clear in this section, which deals with the discrimination between the two enantiomers of racemic salts by chiral macrocyclic ligands.

The concept that forms the basis for this part of crown-ether chemistry was formulated by Cram (Cram and Cram, 1974; Cram *et al.*, 1975; Cram and Cram, 1978) in close analogy with the high degreee of stereospecificity displayed by high-molecular weight enzymes (host) in their reactions with low-molecular weight substrates (guest) in terms of recognition and catalysis. Processes in nature such as selective transport through membranes and highly specific enzymic transformations owe their fast rates and specificity to an ideal fitting of host and guest.

Most of the work on chiral recognition in the ground state deals with salts having chiral, primary alkylammonium cations. Another approach is the chiral discrimination between two enantiomeric anions present as counterions in metal-cation complexes (Lehn *et al.*, 1978). Discrimination between enantiomeric transition states will be dealt with in the next section together with non-chiral mimicry of enzymic catalysis.

The molecular cavity of a crown ether that is formed by the hetero (oxygen) atoms becomes chiral upon incorporation of chiral structure elements. Two different types of such elements have been employed, viz. (*i*) configurationally stable chiral carbon atoms, and (*ii*) structure elements that owe their chirality to restricted rotations about a carbon–carbon bond in the molecule. Crown ethers of the first type were derived from L-proline and D-ψ-ephedrine (Wudl and Gaeta, 1972), cyclohexane-1,2-diols (Hayward *et al.*, 1976; Stoddart and Wheatley, 1974; Burden *et al.*, 1977), L-(+)-tartaric acid (Girodeau *et al.*, 1975; Curtis *et al.*, 1975, 1977a) and various carbohydrates (Laidler and Stoddart, 1977b; Hain *et al.*, 1978). Crown ethers of the second type were used extensively by Cram and his group; they contain one or more 1,1′-dinaphthyl (DD) or 1,1′-ditetralyl (TT) subunits as the chiral element(s) (Kyba *et al.*, 1973a, 1977a; Cram *et al.*, 1978a,b). A few crown ethers have been reported in which both types of chiral elements are present (Curtis *et al.*, 1976). Bicyclic polyethers containing chiral 1,1′-dinaphthyl subunits are also known (Dietrich *et al.*, 1974). Chiral recognition by crown ethers will be reviewed separately for compounds with one and with two chiral elements. Crown ethers with three 1,1′-dinaphthyl subunits do not form complexes with racemic salts (De Jong *et al.*, 1975; Cram *et al.*, 1975).

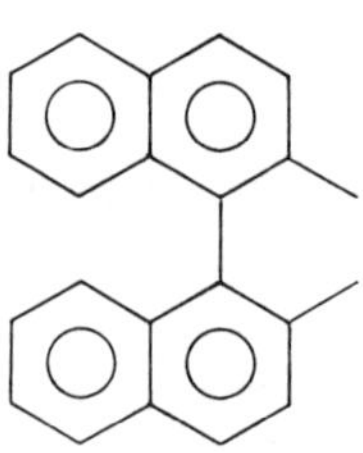

DD

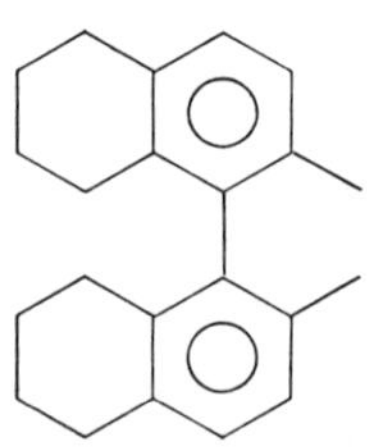

TT

Optical stability is a prerequisite for the use of crown ethers in chiral recognition studies. Cram *et al.* studied the optical stability of (–)-*S*-dinaphthyl-17-crown-5 [265] and found that no racemization occurred upon heating for 6 hours at 205 °C in diethylene glycol. Heating for 202 hours at the same temperature resulted in a loss of 9% of the optical activity (Cram *et al.*, 1978b; Kyba *et al.*, 1977a). Equilibration of diastereoisomeric crown ethers with three, 1,1′-dinaphthyl elements like (*RRR*)(*SSS*)-[266] and (*RRS*)-(*SSR*)-[266] occurred only on melting at 340 °C (7 minutes). A comparative study of the optical stabilities of crown ethers with two 1,1′-dinaphthyl [267] or 1,1′-ditetralyl groups [268] by heating them at 226 °C for 168 hours in diphenyl ether showed that less than 3% racemization had taken place (Cram *et al.*, 1978b). The conclusion is that under the conditions of chiral recognition experiments racemization can be neglected.

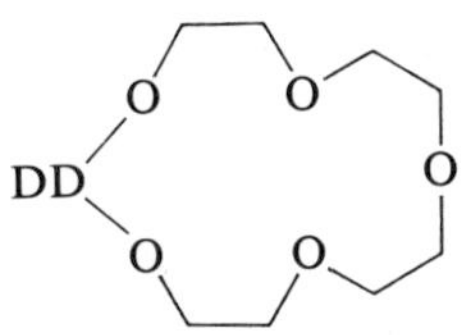

[265]

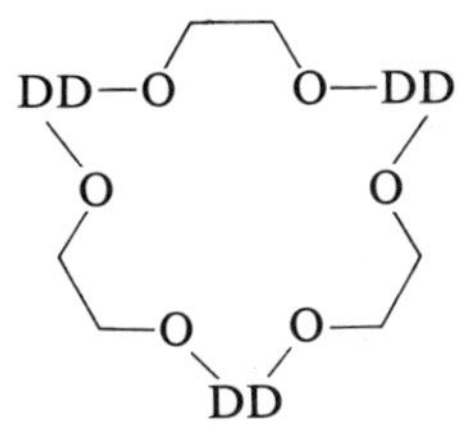

[266]

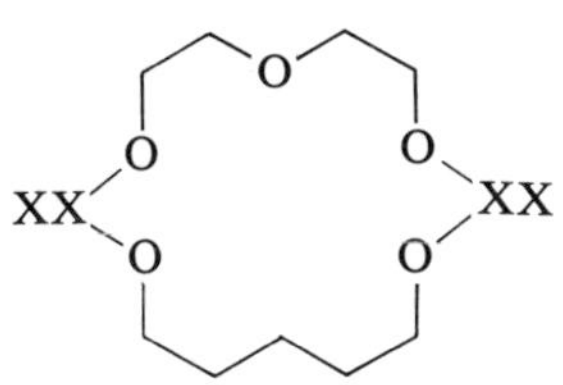

[267], XX = DD

[268], XX = TT

CHIRAL RECOGNITION BY DINAPHTHYL-CROWN ETHERS

Chiral recognition experiments with dinaphthyl-crown ethers have been carried out in two different ways (Helgeson *et al.*, 1973b; Timko *et al.*, 1978):

(*i*) Recognition of racemic host molecules by optically pure guests.

(*ii*) Recognition of racemic guests by optically pure host molecules.

Experiments of the first type were carried out using L-valine as the guest and a series of 1,1′-dinaphthyl-crown ethers bearing side chains with terminal carboxyl or hydroxyl functions [269]–[280]. These substituents are attached to the crown ethers via the 3,3′-(or 6,6′-)positions of the dinaphthyl groups. Chiral recognition was measured by partition experiments in which the components were distributed over a predominantly aqueous phase and a

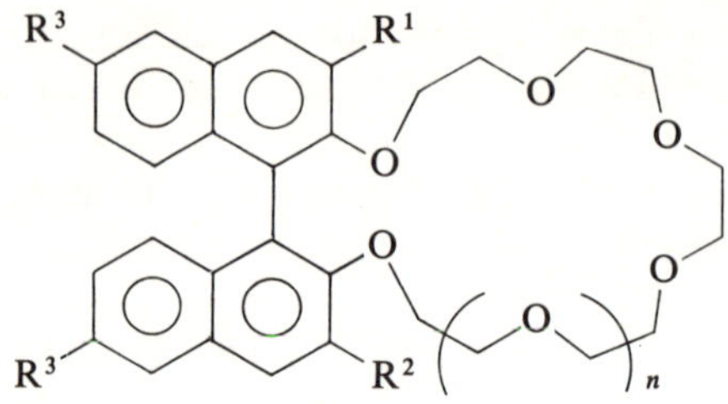

No.	n	R^1	R^2	R^3
[269]	1	$CH_2OCH_2CO_2H$	$CH_2OCH_2CO_2H$	H
[270]	1	$CH_2OCH_2CO_2CH_3$	$CH_2OCH_2CO_2CH_3$	H
[271]	1	H	$CH_2OCH_2CO_2H$	H
[272]	1	CH_3	$CH_2OCH_2CO_2H$	H
[273]	1	H	H	$CH_2OCH_2CO_2H$
[274]	0	$CH_2OCH_2CO_2H$	$CH_2OCH_2CO_2H$	H
[275]	2	$CH_2OCH_2CO_2H$	$CH_2OCH_2CO_2H$	H
[276]	1	$CH_2SCH_2CO_2H$	$CH_2SCH_2CO_2H$	H
[277]	1	$CH_2SCH_2CH_2CO_2H$	$CH_2SCH_2CH_2CO_2H$	H
[278]	1	$CH_2CH_2CO_2H$	$CH_2CH_2CO_2H$	H
[279]	1	CH_2OH	CH_2OH	H
[280]	1	CH_2OH	$CH_2OCH_2CO_2H$	H

chloroform layer. Chiral recognition takes place mainly in the aqueous phase and is expressed as CRF, chiral recognition factor, in this case applicable to the predominantly aqueous phase or as EDC, enantiomer distribution constant, for the host between the two phases. These two parameters are related by the expressions (51) and (52) in which CRF $= [H_A]_{H_2O}/[H_B]_{H_2O}$ and

$$\mathrm{EDC} = K_A/K_B = \frac{[H_A]_{H_2O}}{[H_A]_{CHCl_3}} \times \frac{[H_B]_{CDCl_3}}{[H_B]_{H_2O}} \tag{51}$$

$$\mathrm{EDC} = \mathrm{CRF}\,.\,[H_B]_{CDCl_3}/[H_A]_{CHCl_3} \tag{52}$$

H_A is the more and H_B is the less complexed host enantiomer in the aqueous phase. In the second type of experiment excess of racemic valine and optically pure (*S*)-[269] were distributed between two immiscible phases. In this experiment a 1:1 complex is formed in the non-aqueous phase in which L-valine dominates by an amount of 12.5% (CRF_{CHCl_3} 1.28 and EDC 1.50). In terms of differences in free energy between the two diastereoisomeric complexes this means a difference in $\Delta(\Delta G^\circ)$ of 0.23 kcal mol^{-1} in favour of the (L)-*S*-[269] complex. Similar experiments have been carried out with crown ethers [270]–[280].

TABLE 58

Enantiomer distribution constant (EDC). Estimates in $CDCl_3$-rich layer in the partitioning of an equivalent of racemic host (H) and optically pure L-valine as guest (G) and their complexes between two liquid phases composed of $RCO_2D(H)$, $CDCl_3$, and D_2O[a]

	Solvents/ml				Concn. ratios at equilibrium[b]		Host in D_2O layer[c]				
Compound	CD_3CO_2D	$CDCl_3$	D_2O	T, °C	D_2O layer, [H]/[G]	$CDCl_3$ layer, [G]/[H]	%	Con-fign.	Opt. purity/%	CRF[d]	EDC[d]
[269]	0.60	0.30	0.20	24	0.5	<0.05	50	*S*	27	1.7	2.9
[269]	0.40[e]	0.20	0.05	0	0.45	<0.10	52	*S*	27	1.7	2.9
[269]	0.35	1.00[f]	0.25	24	0.50	<0.10	50	*S*	8.3	1.25	1.6
[270]	0.60	0.20	0.15	24	0.55	0.20	64	*S*	3	1.06	1.1
[271]	0.65	0.20	0.19	24	0.40	<0.10	60	—	—	—	~1.0
[272]	0.60	0.24	0.21	24	0.50	0.26	42	—	—	—	~1.0
[273]	0.60	0.30	0.30	24	0.40	<0.10	39	—	—	—	~1.0
[275]	0.50	0.30	0.25	24	0.42	0.30	25	*S*	1.2	1.02	1.0
[276]	0.40[e]	0.30	0.14	24	0.50	0.30	38	*S*	12	1.27	1.7
[277]	0.60	0.24	0.22	24	0.55	0.31	64	*S*	4	1.08	1.1
[278]	0.60	0.20	0.28	24	0.60	<0.10	60	*S*	4	1.08	1.2
[279]	0.40	0.20	0.15	24	0.38	0.38	38	*S*	6.5	1.14	1.2
[280]	0.50	0.25	0.20	24	0.50	0.20	50	*S*	7	1.15	1.3

[a] Timko *et al.*, 1978

[b] Determined by ^1H nmr integrations

[c] Host was isolated and weighed and rotation was measured. Percentage is based on total isolated material from both layers = 100%

[d] Calculated from optical purity

[e] HCO_2H used in place of CD_3CO_2D

[f] In benzene rather than $CDCl_3$

From the data in Table 58 several conclusions can be drawn about the effect of the structure on the degree of chiral recognition. In all cases the most stable diastereomer has the same configuration, (L)-*S*. The proposed structure of the complex is represented in [281]. Complexation involves formally a proton transfer from one of the carboxyl groups of the crown ether to the amino group of valine. This ammonium group provides the major binding interaction in the complex through three hydrogen bonds (N^+—H—O) with crown-ether oxygen atoms located in a plane perpendicular to the C*–N axis of valine. The second

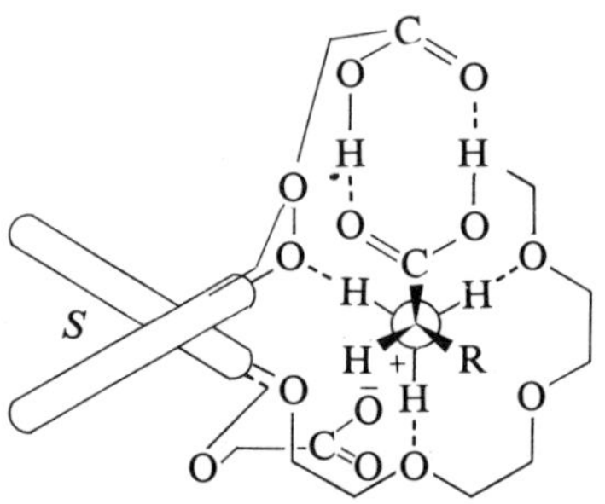

[281], (L)-*S*-[269] complex

carboxyl group of the crown ether is hydrogen bonded to the carboxyl group of valine and this fourth binding interaction provides an enhancement of organization of the complex and restricts the configurational mobility. In the more favourable diastereomeric complex the amino-acid hydrogen atom is located in the chiral cavity close to the 1,1′-dinaphthyl group whereas in the less stable complex (L)-*R*-[269] the positions of the isopropyl group and the hydrogen atom are inverted. This brings the bulky isopropyl group close to the 1,1′-dinaphthyl moiety or alternatively the favourable interaction of the two carboxyl groups is lost (R and COOH inverted in complex [281]).

This generalized model accounts for the observed relationship between structure of the crown ether and the degree of chiral recognition in terms of the following parameters:

Ring size. Only the 20-membered crown ethers display chiral recognition. The 17-membered ring [274], despite having the same side chains as crown ether [269], did not complex valine well enough to allow transfer of the neutral complex to the aqueous phase. The corresponding crown ether with a 23-membered ring [275] complexes valine well (H/G ratio of 0.42) but is incapable of chiral discrimination between the two enantiomeric complexes. Probably complexation takes place through the set of three crown-ether oxygen atoms that is the most remote from the chiral 1,1′-dinaphthyl moiety because the aryl oxygens are the less basic donor sites.

Type and position of the side chains. The most favourable positions of the two side chains are the 3- and 3′-positions because the terminal carboxyl group will be placed under and above the plane of the macro-ring. Conformational mobility and chain length further determine the degree of chiral recognition ([269], [276]–[278]) and omission of one or both carboxyl groups reduces the EDC values ([271], [272], [279], [280]). It was suggested that also the different acidities of the terminal carboxyl groups in [269], [276], [277] and [278] might explain at least partly the observed differences in chiral discrimination. The highest values of CRF and EDC in this series are observed for the crown ether that requires the least proton transfer energy since the pK_a-values of [269] and valine are the most compatible, ΔpK_a −1.3, which corresponds to 0.7 kcal mol^{-1}. For the less acidic carboxyl groups of [276]–[277] a higher loss of energy due to proton transfer will have to be compensated for by the hydrogen bonds between host and guest carboxyl groups. This reduces the energy difference between complexes of type [281] and their less structured diastereomeric complexes with valine in the zwitterionic form. The effect of non-structural type parameters on the relative stabilities of diastereomeric complexes was observed but not accounted for.

Total resolution of [269] into its two enantiomers was achieved by liquid–liquid chromatography through complexation to L-valine adsorbed initially on diatomaceous earth (Timko *et al.*, 1978). On the basis of comparative chromatographic studies, the separation factors (α) and the EDC values were correlated (Cram *et al.*, 1975) (Table 59).

TABLE 59

Correlation between chiral recognition of crown ethers [269], [276] and [278] towards L-valine in extraction (EDC) and chromatographic (α) experiments[a]

Crown ethers	Chiral recognition (EDC)	Separation factor (α)
[269]	2.9	3.1–3.2
[276]	1.7	1.9–2.1
[278]	1.1	1.0

[a] Cram *et al.*, 1975

In the above crown ethers the chiral recognition arises from the chirality of the crown ether cavity and is due to the 1,1′-dinaphthyl structure element. The *degree* of enantiomeric differentiation is substantially enhanced if the conformational mobility of the molecule that is complexed by the crown ether through three N^+–H–O hydrogen bonds is restricted by a fourth binding site like the hydrogen bonding between the host- and guest-carboxyl groups. Other crown ethers in which there are on first sight no such additional binding sites do show chiral recognition. Macrotricyclic polyethers having one chiral

1,1′-dinaphthyl group [282] exhibited enantiomeric differentiation when complexed to racemic ammonium salts like DL-α-naphthylethylammonium chloride both in "one-plate" extraction and in transport through liquid membranes (Lehn *et al.*, 1978; Dietrich *et al.*, 1974). Chiral recognition takes place in the organic phase with a maximum of 15% enantiomeric excess, the (L)-*S*-isomer being the more stable diastereomeric complex. Another crown ether of this type [283] shows chiral recognition in the organic phase with

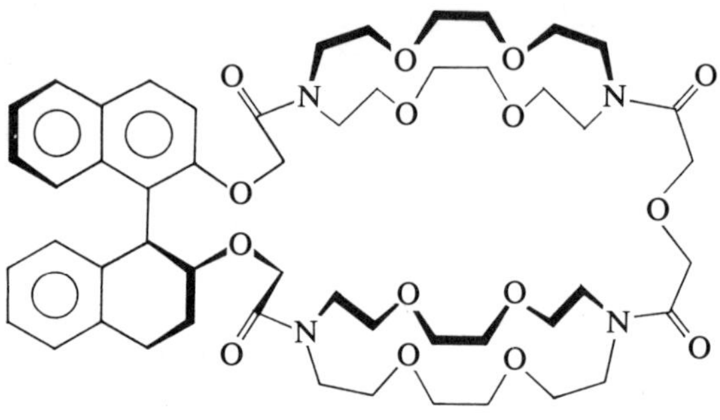

[282]

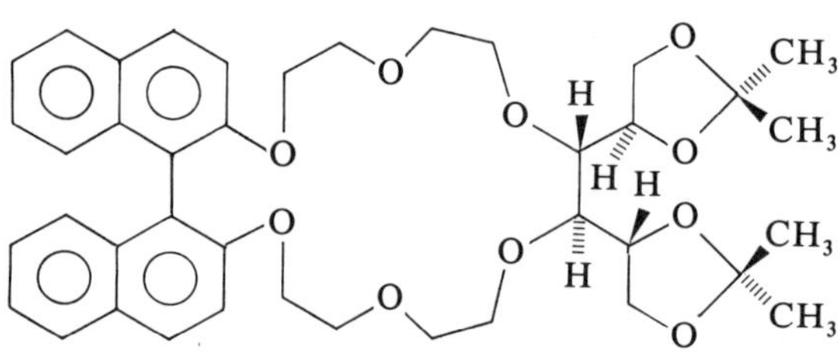

[283]

RS-α-phenylethylammonium hexafluorophosphate. ^{13}C nmr was used to identify the degree of chiral recognition that amounted to an enantiomeric excess (e.e.) value of 2.0 with the *R*-isomer forming the more stable complex with (*R*)-(*D*)-[283] in chloroform. Similarly (*S*)-(*D*)-[283] forms the more stable complex with *S*-α-phenylethylammonium hexafluorophosphate (Curtis *et al.*, 1976).

The results reviewed in this section demonstrate that chiral recognition in complexes between chiral crown ethers and racemic ammonium salts and *vice versa* occurs both in polar and in apolar solvents. The maximum values of 2.9 found for the enantiomeric distribution constants correspond to a difference in free energy of the two diastereomeric complexes of ~0.6 kcal mol^{-1}.

CHIRAL RECOGNITION BY BIS(DINAPHTHYL)-CROWN ETHERS

According to the nomenclature suggested by Cram *et al.* (Kyba *et al.*, 1978) the crown ethers described above are monolocular host molecules because they possess two, usually identical, cavities one on each of the binding faces of the molecule. Following this nomenclature, crown ethers [267] and [268] are dilocular host molecules, having two identical cavities at each of the two binding sites. Compared to all other work on chiral recognition described in this review, the chemistry of these crown ethers [284]–[305] has been investigated most intensively with respect to all the factors that contribute to the degree of enantiomeric differentiation (Kyba *et al.*, 1978; Sousa *et al.*, 1978; Peacock *et al.*, 1978). The results will be reviewed in relation to the most important parameters of the chiral recognition process, viz., host structure, guest structure, effect of the counterion, and non-structural effects. Finally, a generalized model that accounts for both the degree and in most cases for the configurational preference in chiral recognition will be discussed.

Chiral recognition has been determined by three different methods:

(*a*) A ^{1}H nmr method comprising the determination of chemical shifts of the individual diastereomeric complexes and subsequent comparison with the intensities of these absorptions in the spectra of the organic layers (Kyba *et al.*, 1978).

(*b*) Extraction, isolation of guests, and measurement of the optical rotation (Kyba *et al.*, 1978).

(*c*) Measurement of the relative rates of transport of enantiomeric salts through liquid membranes that contain the host molecules (Newcomb *et al.*, 1974). This method is particularly useful if the EDC values are small. From the EDC values obtained, the difference in free energy of pairs of diastereomeric complexes can be calculated [$\Delta(\Delta G°) = \boldsymbol{R}T\ln\mathrm{EDC}$] provided that a number of conditions are fulfilled.

Effect of host structure

The maximum observed free energy difference between two enantiomeric host-guest complexes in which one 1,1′-dinaphthyl element is the only source of chirality in the crown ether is about 0.3 kcal mol^{-1}. Improvement of the free energy difference can be achieved by introduction of two such elements. Unfortunately crown ethers with three 1,1′-dinaphthyl groups did not form complexes with primary ammonium salts (de Jong *et al.*, 1975). The dilocular chiral crown ether [294] forms complexes of different stability with *R*- and *S*-α-phenylethylammonium hexafluorophosphate. The (*R*)-*RR*-[284] complex was the more stable by 0.3 kcal mol^{-1} at 0°C (EDC value 1.77) (Kyba *et al.*, 1973b). Crown ether [284] also discriminates between the two enantiomers of phenylglycine methyl ester hexafluorophosphate and valine methyl ester

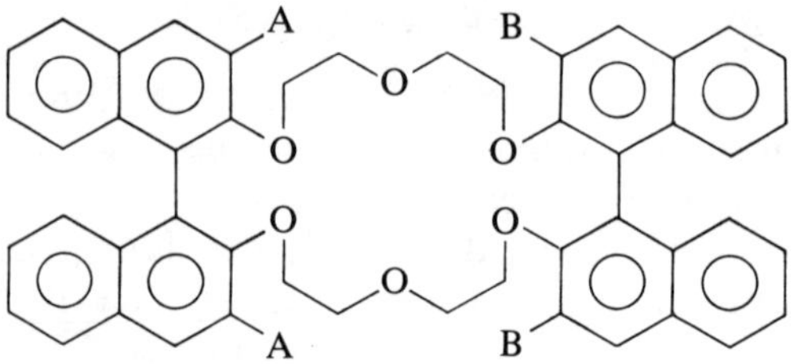

[284], A = B = H

[285], A = CH_3, B = H

286], A = i-Pr, B = H

[287], A = CH_2Cl, B = H

[288], A = CH_2Br, B = H

[289], A = B = CH_3

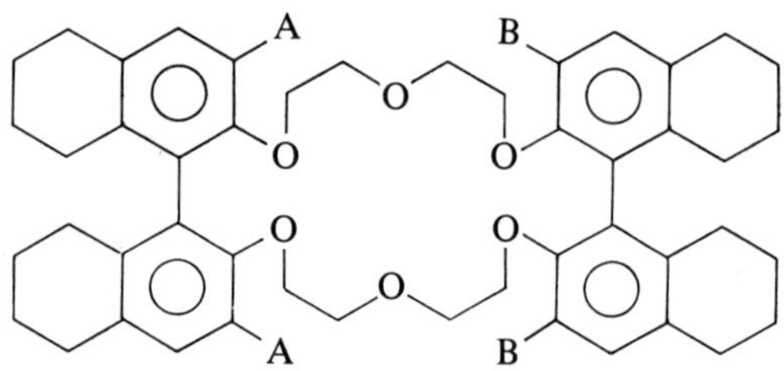

[290], A = B = H

[291], A = CH_3, B = H

[292], A = B = Br

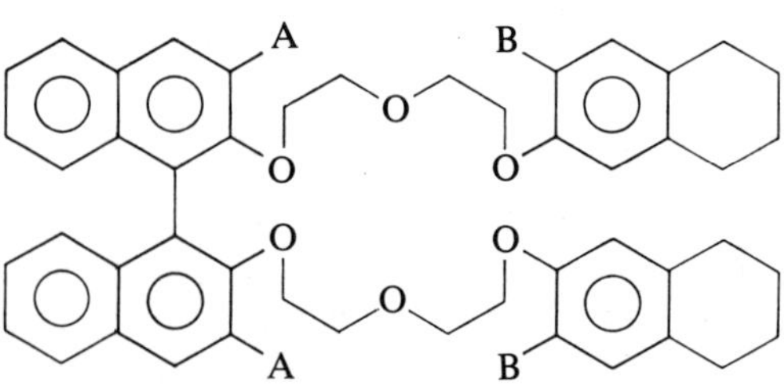

[293], A = B = H

[294], A = CH_3, B = H

[295], A = H, B = CH_3

[296], A = H, B = Br

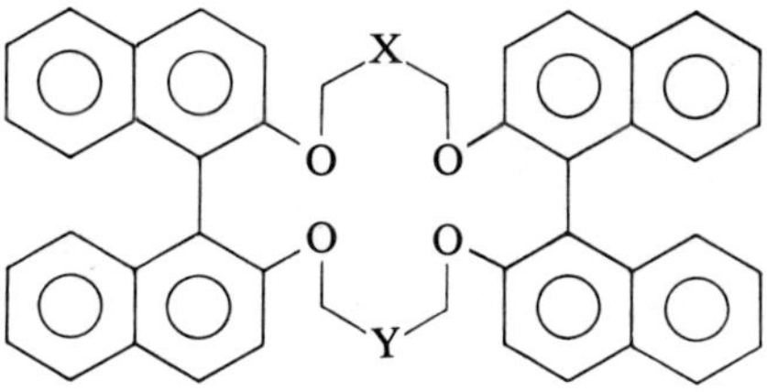

[297], X = CH_2OCH_2, Y = $(CH_2)_3$

[298], X = CH_2OCH_2, Y = 1,3-benzo

[299], X = CH_2OCH_2, Y = 2,6-pyridino

[300], X = Y = 2,6-pyridino

[301], X = Y = 2,6-pyridino, Y = 1,3-benzo

[302], X = 2,6-pyridino, Y = $(CH_2)_3$

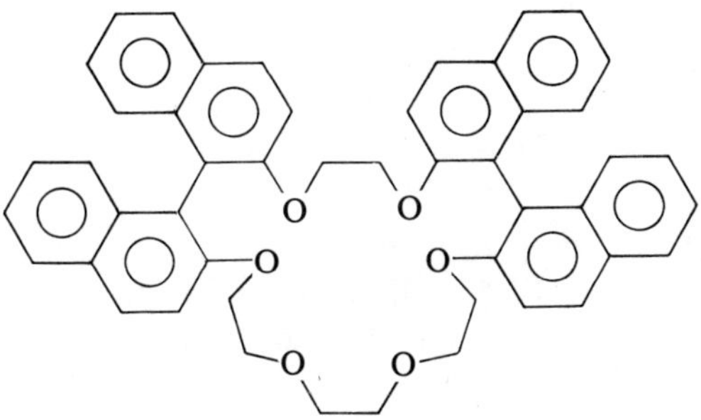

[303]

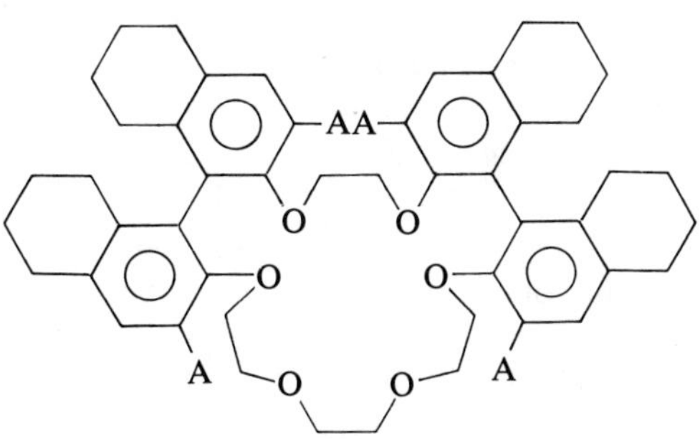

[304], A = H

[305], A = Br

TABLE 60

Chiral recognition between dilocular crown ethers (0.2 M in $CDCl_3$) and $RCH(COOCH_3)NH_3PF_6$ (1.2 M in D_2O/4 M $LiPF_6$)[a]

Crown ether	Temp. /°C	R	G/H	EDC[b]	$\Delta(\Delta G)$ /kcal mol^{-1}
RR-[284]	−18	C_6H_5	1.0	3.1	−0.57
RR-[284]	−10	$CH(CH_3)_2$	0.6	1.5	−0.21
SS-[297]	−15	C_6H_5	~0	—	—
SS-[297]	−6	$CH(CH_3)_2$	~0	—	—
SS-[298]	−10	C_6H_5	~0	—	—
SS-[298]	−10	$CH(CH_3)_2$	~0	—	—
SS-[299]	−14	C_6H_5	1.2	1.7	−0.27
SS-[299]	−16	$CH(CH_3)_2$	0.8	1.24	−0.11
SS-[300]	−16	C_6H_5	0.7	2.0	−0.36
SS-[300]	−16	$CH(CH_3)_2$	0.3	1.3	−0.14
SS-[301]	−17	C_6H_5	0.3	1.0	0
SS-[301]	−16	$CH(CH_3)_2$	~0	—	—
SS-[302]	−13	C_6H_5	0.4	1.35	−0.16
SS-[302]	−16	$CH(CH_3)_2$	0	—	—

[a] Kyba *et al.*, 1978; Gokel *et al.*, 1975a

[b] Except in the case of the complex of [284] with valine methyl ester the *SS*-L or *RR*-D complexes are the most stable

hexafluorophosphate in favour of the (L)-*SS*-[284] complex, with EDC values of 3.1 and 1.5, respectively. This corresponds to free energy differences of 0.57 and 0.21 kcal mol^{-1} (Table 60) (Kyba *et al.*, 1978). In similar crown ethers ([297]–[302]) these values are reduced due to the different nature or the loss of binding sites available for the complexation (Gokel *et al.*, 1975a; Kyba *et al.*, 1978). The general tendency is that complexes with the highest binding constants in a series of similar chiral crown ethers exhibit the highest degree of chiral recognition. The non-symmetrical crown ethers [303] and [304] in which the two chiral elements are connected by one and three ethylene oxide units have, unlike [284], two non-identical chiral cavities on the binding face (Gokel *et al.*, 1975b). Chiral recognition by these compounds and racemic phenylglycine methyl ester or valine methyl ester salts is much lower (Table 61) or even absent. Another difference between these and the symmetrical crown ethers is that the other enantiomer of phenylglycine methyl ester salt is complexed preferentially. Substitution of dinaphthyl for ditetralyl slightly improves the EDC value (compare [284] and [290] or [303] and [304]), but in general the effect of the two different groups is the same (Table 61).

A substantial increase in the enantiomeric differentiation is observed when the 3- and 3′-positions of the 1,1′-dinaphthyl or 1,1′-ditetralyl groups are substituted (Helgeson *et al.*, 1974; Peacock and Cram, 1976; Peacock *et al.*,

TABLE 61

Effect of the symmetry of the crown-ether structure on chiral recognition of $RCH(COOCH_3)NH_3X$ at 0°C[a]

Crown ether	R	X^-	EDC	$\Delta(\Delta G°)$ /kcal mol^{-1}	More stable complex
[284]	C_6H_5	ClO_4^-	2.4	−0.48	*SS*-L
[284]	C_6H_5	PF_6^-	2.8	−0.56	*RR*-D
[284]	$CH(CH_3)_2$	ClO_4^-	1.1	−0.05	*SS*-D
[290]	C_6H_5	ClO_4^-	3.1	−0.62	*RR*-D
[290]	$CH(CH_3)_2$	ClO_4^-	1.2	−0.10	*RR*-D
[303]	C_6H_5	ClO_4^-	1.8	−0.32	*SS*-D
[303]	$CH(CH_3)_2$	ClO_4^-	1.3	−0.14	*SS*-D
[304]	C_6H_5	ClO_4^-	1.1	−0.05	*SS*-D
[305]	C_6H_5	ClO_4^-	1.8	−0.22	*SS*-D

[a] Peacock *et al.*, 1978

1978) (Table 62). Compared with the unsubstituted crown ether [284] the largest differences exist with [285] in which one of the dinaphthyl groups is substituted at the 3- and 3′-positions with a methyl group [$\Delta(\Delta G°)$-values up to −1.9 kcal mol^{-1}]. Similarly, the presence of two bromine atoms in these positions gives an increase to −1.25 kcal mol^{-1}. The larger isopropyl group has a smaller effect and [286] is better than [284] with only 0.26 kcal mol^{-1}. The effect of two chloromethyl groups is less clear; in "one-plate" extraction experiments [287] and [288] were reported to be poor complexing ligands

TABLE 62

Effect of substituents in *bis*(dinaphthyl)-22-crown-6 at the 3,3′-positions on chiral recognition of $RCH(COOCH_3)NH_3X$[a]

Crown ether	Temp. /°C	R	X^-	R	$\Delta(\Delta G°)$ /kcal mol^{-1}	More stable complex
[285]	0	C_6H_5	PF_6^-	31	−1.90	*RR*-D
[285]	0	C_6H_5	ClO_4^-	21	−1.65	*RR*-D
[294]	0	C_6H_5	PF_6^-	31	−1.90	*RR*-D
[295]	0	C_6H_5	PF_6^-	20	−1.60	*RR*-D
[291]	0	C_6H_5	PF_6^-	13.6	−1.40	*RR*-D
[291]	0	C_6H_5	ClO_4^-	10.2	−1.30	*SS*-L
[296]	0	C_6H_5	PF_6^-	11.5	−1.25	*RR*-D
[296]	−10	C_6H_5	PF_6^-	4.8	−0.82	*RR*-D
[292]	0	C_6H_5	ClO_4^-	−1.5	−0.22	*SS*-L
[286]	0	C_6H_5	PF_6^-	5.0	−0.82	*RR*-D

[a] Peacock *et al.*, 1978

(Peacock *et al.*, 1978), but in experiments based on kinetic resolution through liquid membranes a high rate difference, usually corresponding to a large EDC value, for [287] with phenylglycine methyl ester salt is mentioned (Newcomb *et al.*, 1974). Substitution of both chiral elements in crown ethers of this type with methyl groups [289] and/or bromine groups ([292], [296] and [305]) reduces both the complexation constant of the complexes with ammonium salts and the chiral recognition (Table 62). It is clear that two competing factors contribute to the overall effect of complexation. The maximum improvement relative to the unsubstituted crown ether [284] is reached with two methyl groups in one of the two chiral elements. Further increase of the size of substituents in the 3,3′-positions or a similar type of substitution in the second chiral element decreases the enantiomeric differentiation. This might be due to a decrease in absolute binding constants of the complexes of these crown ethers and the ammonium salts. The magnitude of the effect of two methyl groups on the chiral recognition of five different amino acid methyl ester hexafluorophosphates is between -0.7 and -1.3 kcal mol^{-1} (Table 63).

TABLE 63

Effect of two methyl groups in the 3- and 3′-positions ([284] compared with [285]) on the free energy differences between diastereomeric complexes[a]

R in $RCH(COOCH_3)NH_3PF_6$	$\Delta(\Delta G°)$ [285] − $\Delta(\Delta G°)$ [284] /kcal mol^{-1}
C_6H_5	−1.3
p-HOC_6H_4	−0.7
$(CH_3)_2CH$	−1.1
$C_6H_5CH_2$	−1.2
$CH_3SCH_2CH_2$	−0.7

[a] In $CDCl_3$ at 0 to −10°C; Peacock *et al.*, 1978; Peacock and Cram, 1976

Effect of guest structure

Extractions of aqueous solutions of racemic amino-acid ester salts with solutions of *RR-bis*(dinaphthyl)-22-crown-6 [284] in chloroform revealed the dependence of the enantiomeric distribution constant on the structure of the amino acid ester (Table 64). In order to limit the concentrations of complex in the aqueous phase, inorganic salts were added. In the case of tyrosine, serine and alanine no extraction of salt was observed; obviously these salts form very hydrophilic complexes. The highest degree of chiral recognition was found with [284] and *p*-hydroxyphenylglycine methyl ester hexafluorophosphate [$\Delta(\Delta G°)$

TABLE 64

EDC values for the extraction of racemic guests $RCH(COOR')NH_3PF_6$ (1.2 M in D_2O, 4 M $LiPF_6$) by crown ether *RR*-[284] (0.20 M in $CDCl_3$)[a]

Temp. /°C	R	R′	G/H	Most stable complex	EDC	$\Delta(\Delta G°)$ /kcal mol^{-1}
−18	C_6H_5	CH_3	0.98	*RR*-D	3.1	−0.57
25	C_6H_5	$CH(CH_3)_2$	0.77	*SS*-L	4.0	−0.82
0	C_6H_5	$C(CH_3)_3$	0.70	*SS*-L	4.4	−0.81
−6	*p*-HOC_6H_4	CH_3	0.50	*RR*-D	3.4	−0.68
−16	*p*-HOC_6H_4	CH_3	0.66	*RR*-D	5.0	−0.82
0	*p*-$CH_3OOC(C_6H_4)$[b]	CH_3[b]	—	*RR*-D	2.3	−0.46
0	*p*-$Cl(C_6H_4)$	CH_3[b]	—	*RR*-D	1.3	−0.15
−10	$(CH_3)_2CH$	CH_3	0.63	*RR*-L	1.5	−0.21
−10	$C_6H_5CH_2$	CH_3	0.73	*RR*-L	1.8	−0.32
−5	$CH_3SCH_2CH_2$	CH_3	1.00	*RR*-L	1.7	−0.28
25	$(CH_3)_2CH$	$CH(CH_3)_2$	0.70	—	1.0	0

[a] Kyba *et al.*, 1978
[b] In $CDCl_3$:CD_3CN (9:1); ClO_4^- replaces PF_6^- (Peacock *et al.*, 1978)

−0.82 kcal mol^{-1}]. In the case of phenylglycine esters the type of ester influenced the EDC value (t-Bu ~ i-Pr > Me). This enantiomeric differentiation between two enantiomers in a "one-plate" extraction was extended by Cram *et al.* (Sousa *et al.*, 1978; Sousa *et al.*, 1974) to complete optical resolution of racemic amines and amino-acid esters. In the first type of experiments the stationary aqueous phase consists of solutions of racemic amino-acid ester salts adsorbed on silica gel, while the mobile phase consists of chiral crown ether [284] in chloroform. The results, given in Table 65, reveal that the EDC

TABLE 65

Liquid–liquid chromatographic runs with *RR*-[284] in $CHCl_3$ in the mobile phase and racemic guest in the aqueous stationary phase [a]

Temp. /°C	RNH_3^+ (PF_6^-)	Separation factor, α	EDC (K_A/K_B)	Configuration of the more stable complex
0	$C_6H_5CH(CH_3)NH_3^+$	1.8	1.8	*RR*-S
25	$C_6H_5CH(CH_3)NH_3^+$	1.5	1.5	*RR*-S
−13	$C_6H_5CH(CO_2CH_3)NH_3^+$	2.5	2.5	*RR*-D
−15	*p*-$HOC_6H_4CH(CO_2CH_3)NH_3^+$	3.6	5	*RR*-D
−15	*p*-$HOC_6H_4CH(CO_2CH_3)NH_3^+$	2.4[b]	—	*RR*-D

[a] Sousa *et al.*, 1978
[b] CH_2Cl_2 was substituted for $CHCl_3$

values obtained from "one-plate" extractions nicely correspond to the separation factor α provided that equilibrium is reached in the elution process (Sousa *et al.*, 1978). In some cases a base-line separation of the two enantiomeric complexes was achieved.

In the second type of chromatographic experiments chiral crown ethers of type [284] covalently attached to silica were equilibrated with solutions of racemic ammonium salts or amino-acid ester salts solubilized by addition of 18-crown-6 or alcohols in chloroform or dichloromethane (Table 66). The

TABLE 66

Optical resolution at 25°C of amine salts by solid–liquid chromatography on a silica gel column containing covalently bound (*RR*)-host of type [284][a]

G(R^*NH_3X)	Mobile phase	H/G[b]	Separation factor, α	Configuration of the more stable complex
$C_6H_5CH(CH_3)NH_3PF_6$	CH_2Cl_2	21	1.5	*RR*-S
$C_6H_5CH(CH_3)NH_3PF_6$	$CHCl_3$	84	1.7	*RR*-S
$C_6H_5CH(CO_2CH_3)NH_3Cl$	CH_2Cl_2	20	1.2	*RR*-D
$C_6H_5CH(CO_2CH_3)NH_3PF_6$	$CHCl_3$	50	1.4	*RR*-D
$C_6H_5CH(CO_2CH_3)NH_3PF_6$	$CHCl_3$	128	1.6	*RR*-D
$C_6H_5CH(CO_2CH(CH_3)_2)NH_3Cl$	CH_2Cl_2	80	2.9	*RR*-D
p-$HOC_6H_4CH(CO_2CH_3)NH_3Cl$	$CHCl_3$	70	6.4	*RR*-D
$(CH_3)_2CHCH(CO_2CH_3)NH_3Cl$	$CHCl_3$	23	1.7	*RR*-L
$C_6H_5CH_2(CO_2CH_3)NH_3Cl$	$CHCl_3$	23	4.4	*RR*-L
p-$HOC_6H_4CH_2CH(CO_2CH_3)NH_3Cl$	$CHCl_3$	82	2.3	*RR*-L
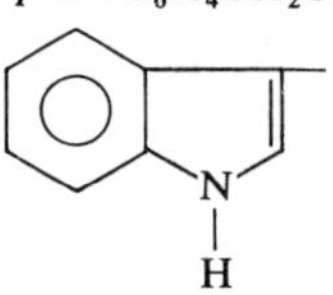 $CH_2CH(CO_2CH_3)NH_3Cl$	$CHCl_3$	70	6.4	*RR*-D

[a] Sousa *et al.*, 1978; Sogah and Cram, 1975
[b] Moles of host to moles of guest

results are in line with those obtained from extraction experiments (Table 64) with respect to both relative EDC (or α) values and the configuration of the more stable complex. The effect of the nature of the substituent in amino-acid esters on the degree of chiral recognition is also observed in experiments with crown ethers [285] and [290] (Table 67). The following general tendencies emerge from these data:

(*i*) Aryl groups generally give rise to higher enantiomeric differentiation.

(*ii*) The influence of *p*-substitutents in the aryl groups can not be accounted for by steric effects but is related to the electronic effect exerted on the phenyl ring and on the amino-ring ester as a whole.

TABLE 67

Effects of group (R) on the chiral recognition of racemic amino-acid esters by [285] and [290] at 0°C[a]

Crown ether	$RCH(COOCH_3)NH_3^+$	X^-	EDC[c]	$\Delta(\Delta G°)$ /kcal mol^{-1}
[285]	C_6H_5	PF_6^-	31.0	−1.90
[285]	p-HOC_6H_4	PF_6^-	8.9[b]	−1.30
[285]	$(CH_3)_2CH$	PF_6^-	5.3	−0.87
[285]	$CH_3SCH_2CH_2$	PF_6^-	2.2	−0.42
[285]	$C_6H_5CH_2$	PF_6^-	5.3	−0.87
[290]	C_6H_5	ClO_4^-	3.2	−0.63
[290]	p-HOC_6H_4	ClO_4^-	6.6	−1.00
[290]	p-$CH_3OOCC_6H_4$	ClO_4^-	3.6	−0.63
[290]	p-ClC_6H_4	ClO_4^-	2.1	−0.40

[a] Peacock *et al.*, 1978

[b] A value of 38 has also been reported (Peacock and Cram, 1976)

[c] In all cases the *RR*-D configuration was the most stable

(*iii*) The configuration of the most stable enantiomeric complex is the same for the entire series.

The response of the *bis*(ditetralyl)-crown ether [290] and the bis(dinaphthyl)-crown ether [284] (see Table 64) to the changes in the guest structure is very similar: HO > CH_3OOC ~ H > Cl. For the 3,3′-dimethylbis(dinaphthyl)-crown ether [285] this order seems to be reversed: H > OH with Ar > t-Bu > $C_6H_5CH_2$ > $CH_3SCH_2CH_2$ (Peacock and Cram, 1976; Helgeson *et al.*, 1974; Peacock *et al.*, 1978). More data on the effect of substituents in aryl groups of amino-acid (ester) salts were obtained from multi-plate equilibration with 3,3′-dimethylbis(dinaphthyl)-22-crown-6 covalently bound to macroreticular cross-linked polystyrene resin via the 6- and 6′-positions (Sogah and Cram, 1976, 1979). Resolution factors (Table 68) ranged from 1.4 to 26, corresponding to $\Delta(\Delta G°)$ values for the diastereomeric complexes of −0.18 to −1.77 kcal mol^{-1}, and base-line separation could be achieved. Generally the configuration of the more stable complex was the same. In the series of substituted phenylglycine esters a linear free-energy relationship between substituent effect and difference in free energy $\Delta(\Delta G°)$ was observed for p-HO, p-H, p-CH_3OOC and p-Cl substituents ($\rho = -0.54$).

Most of the work on chiral recognition has been carried out with amino-acid ester salts. The limited number of available data of similar experiments with amino-*acid* salts point to an increased enantiomeric differentiation (Table 69; Peacock and Cram, 1976). The reason might be the higher π-acidity of the carboxyl group (*vide infra*). In solid–liquid chromatography (see Table 68) the difference between amino-acid salts and amino-acid ester salts is not consistent and not always in favour of the amino-acid salts. However, it should be

TABLE 68

Optical resolution of $RCH(COOR')NH_3ClO_4$ by solid–liquid chromatography using *RR*-[285] covalently attached to polystyrene[a]

R	R′	% CH_3CN in $CDCl_3$ (w/w)	Separation factor, α	$\Delta(\Delta G°)$ (kcal/mol)	Configuration of the more stable complex
C_6H_5	H	10	24.0	1.726	*RR-R*
p-HOC_6H_4	H	10	6.1	0.982	*RR-R*
p-$HOC_6H_4CH_2$	H	10	1.9	0.349	*RR-R*
$C_6H_5CH_2$	H	4	2.3	0.453	*RR-R*[b]
$C_6H_8NCH_2$[c]	H	20	6.1	0.982	*RR-R*
$(CH_3)_2CH$	H	10	2.3	0.452	*RR-R*
$C_2H_5(CH_3)CH$	H	5	1.9	0.349	*RR-R*
$(CH_3)_3C$	H	5	1.9	0.349	*RR-R*
CH_3	H	4	1.5	0.220	*RR-R*
$CH_3SCH_2CH_2$	H	4	1.4	0.183	*RR-R*
p-HOC_6H_4	CH_3	10	26.0	1.770	*RR-R*
C_6H_5	CH_3	10	18.5	1.585	*RR-R*
p-$CH_3OOCC_6H_4$	CH_3	10	12.6	1.376	*RR-R*
p-ClC_6H_4	CH_3	10	8.5	1.163	*RR-R*
p-$FC_6H_4CH_2$	CH_3	10	8.5	1.163	*RR-S*[d]
$C_6H_5CH_2$	CH_3	10	6.4	1.008	*RR-R*[b]
p-$HOC_6H_4CH_2$	CH_3	10	4.7	0.841	*RR-R*

[a] Sogah and Cram, 1976
[b] Sogah and Cram, 1979
[c] Tryptophan side chain
[d] Not mentioned in Sogah and Cram, 1979

TABLE 69

Enantiomeric differentiation of amino acid and the corresponding amino-acid methyl ester salts $RCH(COOR')NH_3X$ by crown ether [285][a]

R	R′	X^-	% CD_3CN in $CDCl_3$ (w/w)	Ratio H/G	EDC[b]	$\Delta(\Delta G°)$ /kcal mol^{-1}
C_6H_5	H	ClO_4^-	23.1	0.92	52	−2.15
C_6H_5	CH_3	ClO_4^-	0	0.91	21	−1.65
p-$HO(C_6H_4)$	H	ClO_4^-	23.1	0.80	48	−2.10
p-$HO(C_6H_4)$	CH_3	ClO_4^-	0	0.72	25	−1.75
$(CH_3)_2CH$	H	ClO_4^-	23.1	0.68	11	−1.30
$(CH_3)_2CH$	CH_3	PF_6^-	0	0.75	5.3	−0.87
$CH_3S(CH_2)_2$	H	ClO_4^-	23.1	0.82	3.2	−0.63
$CH_3S(CH_2)_2$	CH_3	PF_6^-	0	0.95	2.2	−0.42
$C_6H_5CH_2$	H	ClO_4^-	23.1	0.89	3.5	−0.68
$C_6H_5CH_2$	CH_3	PF_6^-	0	1.0	2.3	−0.45

[a] Peacock and Cram, 1976
[b] Except for R = C_6H_5—CH_2 the *RR*-D complexes are the more stable

mentioned that in these experiments the ratio of acetonitrile and chloroform (1 : 9) is lower than that employed in the standard extraction experiments.

Effect of the counterion

The influence of the counterion on the stability of crown-ether complexes in general was reviewed in detail in one of the preceding sections. There it was shown to be an important parameter. The nature of the counterion in diastereomeric complexes of chiral crown ethers with primary ammonium salts also influences the chiral recognition. First of all it greatly determines whether salt can be extracted into the organic phase where the chiral discrimination takes place. In a series of experiments (Kyba *et al.*, 1978) it was shown that when *SS-bis*(dinaphthyl)-22-crown-6 [284] in chloroform was equilibrated with racemic α-phenylethylammonium salts the type of anion also influences the degree of enantiomeric differentiation (Table 70). The highest

TABLE 70

The effect of various ions on the guest/host ratios and EDC-values observed at 0°C in the extraction of racemic α-phenylethylammonium salts by (*SS*)-[284][a]

Initial D_2O phase					
Amine salt			In $CDCl_3$ at equil.	CRF in	
Anion	Conc./M	Inorg. salt	[G]/[H]	$CDCl_3$	EDC
F^-	1	None	0.0	—	—
Cl^-	1	None	0.0	—	—
Br^-	1	None	0.0	—	—
PF_6^-	1	None	1.0	1.8	2.0
PF_6^-	0.4	None	0.6	1.8	2.1
PF_6^-	0.4	NaBr	0.6	1.7	2.0
PF_6^-	0.4	Kbr	0.0	—	—
PF_6^-	1	KPF_6	1.0	1.6	1.7
PF_6^-	1	LiI	1.2	~1.1	1.1
PF_6^-	1	NaI	1.5	~1.0	1.0
PF_6^-	1	KI	1.2	~1.2	1.2
F^-	1	$LiPF_6$	1.0	1.6	1.7
F^-	1	$NaPF_6$	1.0	2.0	2.2
F^-	1	KPF_6	0.8	1.7	1.8
Br^-	1	$LiPF_6$	1.0	1.7	1.9
Br^-	1	$NaPF_6$	0.9	1.7	1.8
Br^-	1	KPF_6	0.5	2	2.1
Br^-	1	$NaAsF_6$	1.0	1.9	2.1
Br^-	1	$NaSbF_6$	1.0	1.75	1.9

[a] Kyba *et al.*, 1978

EDC values were obtained with large, bulky ("soft") anions such as PF_6^-, ClO_4^-, AsF_6^- and SbF_6^-. In these anions the charge is highly delocalized and their tendency to form hydrogen bonds is low. Therefore in the crown-ether complexes with ammonium ions these anions hardly compete with the crown ether oxygen co-ordination sites for the binding to the RNH_3^+ group. This will result in a high degree of structuring of these complexes, a prerequisite for a good differentiation between two enantiomeric cations. With "hard" anions (Cl^-, Br^-) no extraction takes place and consequently their effect on the EDC value is unknown. Anions that occupy positions between the "soft" and "hard" types such as I^- and SCN^- give corresponding complexes that are transferred to the organic phase. Chiral recognition, however, is close to zero probably because these anions form hydrogen bonds with the ammonium group.

TABLE 71

Effect of counter ion on chiral recognition of $C_6H_5CH(CO_2CH_3)NH_3X$[a]

Crown ether	X^-	$-\Delta(\Delta G°)$ /kcal mol^{-1}
[284]	PF_6^-	−0.56
[284]	ClO_4^-	−0.48
[285]	PF_6^-	−1.9
[285]	ClO_4^-	−1.7
[291]	PF_6^-	−1.4
[291]	ClO_4^-	−1.3

[a] Peacock *et al.*, 1978

In view of these results the effect of anions on the chiral resolution by liquid–solid chromatography of racemic amino-acid ester salts on covalently bound chiral crown ethers of type [284] is unexpected (Sousa *et al.*, 1978). The highest resolution factors corresponding to high EDC values were obtained with tyrosine methyl ester hydrochloride and tryptophan methyl ester hydrochloride (see Table 66). This result is attributed to the strong binding of the chloride anions to residual silica hydroxyl groups leaving the cations free to form highly structured complexes.

Table 71 shows the general tendency of a higher chiral recognition in complexes with hexafluorophosphate than with perchlorate anions. This might reflect the slightly higher stability of PF_6^- complexes.

Non-structural effects

Most of these chiral recognition experiments with amino-acid ester salts were carried out in two-phase systems in which the organic solvents were chloroform, dichloromethane or in some cases 1,2-dichlorobenzene (Kyba *et al.*, 1978). With a view to obtaining a sufficiently high complex concentration in the organic phase or to reducing the amount of crown ether in the aqueous phase, inorganic salt is added (Peacock *et al.*, 1978) or acetonitrile (up to 10%) is added to the chloroform. Only small differences exist in the degree of chiral differentiation between the two enantiomers and no changes in the configuration of the more stable complex are observed. In all the experiments, whether one-plate or chromatographic, the EDC values are highest with chloroform (Sousa *et al.*, 1978).

With *amino-acid* salts the effect of the medium is considerably larger. Peacock and Cram (1976) reported that the degree of chiral recognition of DL-phenylglycine perchlorate by crown ether [285] depends on the ratio of acetonitrile and chloroform. The observed EDC values vary from 6 to 52, which corresponds to a difference in free energy of -1.15 kcal mol^{-1} (Table 72). The optimum is very sharply defined (23.1% of acetonitrile) and is

TABLE 72

Effect of the polarity of the medium on the degree of chiral recognition of DL-phenylglycine perchlorate by [285] at 0°C[a]

% CD_3CN in $CDCl_3$ (w/w)	Ratio G/H in org. phase	EDC	$\Delta(\Delta G°)$ /kcal mol^{-1}	Configuration of more stable complex
0	0	—	—	—
9.1	0.15	6.15	−0.99	*RR*-D
16.7	0.58	9.32	−1.21	*RR*-D
23.1	0.92	52.00	−2.15	*RR*-D
28.6	1.00	6.15	−0.99	*RR*-D
33.3	1.00	6.25	−1.00	*RR*-D

[a] Peacock and Cram, 1976

explained in terms of a close balance between displacement of the anion by solvent molecules and competition between RNH_3^+ and CH_3CN protons for the crown ether binding (oxygen) sites. In similar experiments with the corresponding methyl esters no noticeable effect of the amount of acetonitrile in chloroform on the EDC value was observed.

An important factor is the *temperature*. Firstly, more of the salt is extracted into the organic layer at lower temperature (Sousa *et al.*, 1978). This indicates that for these salts the entropy of complexation has a substantial negative value (Kyba *et al.*, 1978). Secondly, the temperature has an effect on the value of the EDC; lower temperatures give rise to higher EDC and $\Delta(\Delta G°)$ values (Table 73). The available data on the effect of temperature hardly allow a quantitative

TABLE 73

Correlation of structures of diastereomeric complexes with estimated enthalpic and entropic contributions to free energy of complexation (0°C)[a]

	$RCH(R')NH_3PF_6$		Parameters/kcal mol^{-1}		
Crown ether	R	R′	$\Delta(\Delta G°)$	$\Delta(\Delta H°)$	$-T\Delta(\Delta S°)$
[285]	C_6H_5	CO_2CH_3	−1.9	−6	4
[285]	p-HOC_6H_4	CO_2CH_3	−1.4	−2.5	1
[296]	C_6H_5	CO_2CH_3	−1.25	−1.6	0.3
[284]	C_6H_5	CO_2CH_3	−0.55	−0.7	0.2
[284]	C_6H_5	CH_3	−0.31	−0.9	0.6

[a] Peacock *et al.*, 1978

evaluation of the free-energy differences into enthalpic and entropic contributions, particularly because these $\Delta(\Delta G°)$ values are obtained from partition experiments rather than from direct measurements.

The data presented in Table 73, however, suggest that the most stable diastereomeric complex owes its stability to stronger binding [higher $-\Delta(\Delta H°)$ contribution] at the cost of much higher degree of organization [$-T\Delta(\Delta S°)$ large, positive]. A preliminary conclusion (Peacock *et al.*, 1978) was that for high chiral recognition the enthalpic and entropic terms should be opposite.

Generalized model for chiral recognition

In crown-ether complexes of the monolocular type like [281] the crown ethers are negatively charged species in which one carboxyl group formally transfers a proton to the amino group. A *second* non-ionized carboxyl group is available for further structuring of the complex; if this is absent the chiral recognition is low. This second carboxyl group provides a fourth binding site in addition to the three O—H—N$^+$ hydrogen bonds. In complexes of the dilocular type such an additional binding site is absent and yet in many cases the chiral recognition is higher. Structure [306] represents such a complex for *SS*-[284], in which binding arises from the three N$^+$—H—O hydrogen bonds. The two smaller groups, S, M occupy one of the two chiral cavities formed by the dinaphthyl

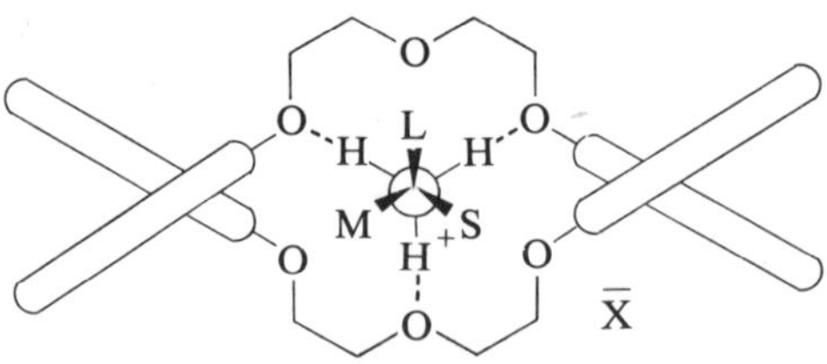

[306]

groups whereas the large group (L) is located in the other one. In the diastereomeric complex the positions of L and M are inverted although this picture might be too simple. Since the crown ether possesses a C_2-axis of symmetry, identical complexes are formed by binding of the ammonium cation on either one face of the crown ethers. The larger the difference in stability between the two diastereomeric complexes, the greater the chiral discrimination will be. This is a purely steric model and, if valid, the direction of the configurational preference could be predicted by using CPK molecular models, provided that the steric requirements of the large and medium-sized substituents are sufficiently different. If additional binding is possible that stabilizes one complex more than the other, the difference in stability will be enhanced or diminished accordingly and the configurational preference will be less predictable. Most of the discussion on a generally applicable structural model by which the direction of the configurational preference could be predicted centres on the nature of the additional fourth binding site and is based on the relationship between the host and guest structures and the EDC [or $\Delta(\Delta G°)$ values]. The model was substantiated by Cram *et al.*, who compared the ^{1}H nmr spectra with the results of an x-ray structure determination of a crystalline complex of D-phenylglycine methyl ester hexafluorophosphate and *SS*-[284] (Kyba *et al.*, 1978; Goldberg, 1977).

Cram *et al.* distinguish five classes of crown ethers of the dilocular type (I–V) and arrive at the following conclusions (Peacock *et al.*, 1978) about the *configurational* preference:

(*a*) In complexes of type I and II with $ArCH(COOCH_3)NH_3X$ the *RR*-(D) or *SS*-(L) diastereomeric form is the most stable.

(*b*) In complexes of type II with $RCH(COOCH_3)NH_3X$ [R = $(CH_3)_2CH$, $CH_3SCH_2CH_2$, (p-X–C_6H_4)CH_2] the *RR*-(D) or *SS*-(L) diastereomers are the most stable, with an exception (Table 68) in chromatographic experiments (Sogah and Cram, 1976) (R = p-F–$C_6H_4CH_2$). In similar complexes formed with crown ethers I the *RR*-(L) or *SS*-(D) diastereomers are the more stable.

(*c*) Complexes of type III are more stable in the *RR*-(D) or *SS*-(L) form.

(*d*) Crown ethers of types IV and V show the same configurational preference, the *RR*-(L) or *SS*-(D) diastereomers being the more stable complexes.

The degree of chiral recognition also differs. Complexes of type I are stable but the structural effects are not sufficiently opposite to give high discrimination between the two enantiomeric salts. Complexes of type II are relatively stable provided that the substituents that extend the chiral barrier are not too bulky. These complexes exhibit the highest chiral recognition, particularly when the groups that extend the chiral barrier are methyl groups. The chiral recognition of types III to V is lower. Information about the structure of the diastereomeric complexes of crown ethers *SS*-[284] and

I; [284], [290], [285], [297]–[302]

II; [286]–[288], [291], [294]–[296]

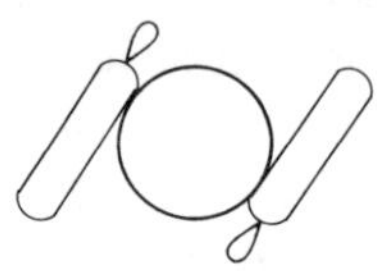

III; [289], [292]

IV; [303], [304]

V; [305]

SS-[285] with D- and L-phenylglycine methyl ester perchlorate was obtained from ^{1}H nmr spectroscopy, and the differences in chemical shift between protons in the diastereomeric complexes could be explained on the basis of structure [306]. Single crystal x-ray analysis of the *less* stable diastereomeric complex (D)-*SS*-[284] (Goldberg, 1977; Kyba *et al.*, 1978) confirms the postulated structure of this complex. It reveals the following important information about the structure (see Fig. 4).

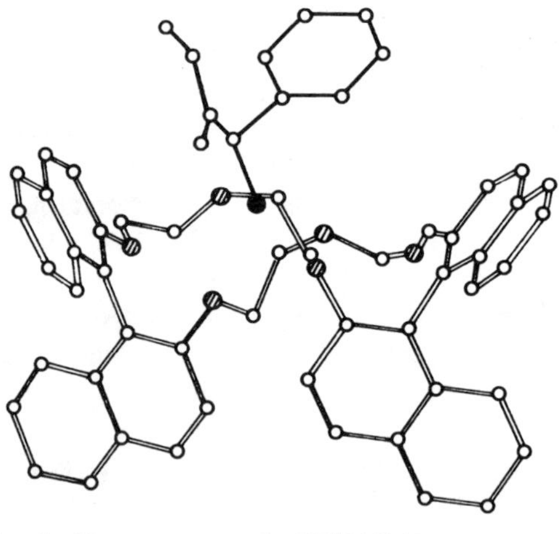

FIG. 4 X-ray structure of D-*SS*-[284] (Goldberg, 1977)

(*a*) The major binding contribution comes from a tripod N^{+}—H—O type of hydrogen bonding.

(*b*) The C*—N axis is perpendicular to the plane of the three crown ether oxygen atoms that are hydrogen bonded.

(*c*) The ester group is located in one cavity, the hydrogen and phenyl groups are located in the other.

(*d*) The plane of the ester group, *and not of the aryl group*, is parallel to the naphthalene ring.

(*e*) The NCH hydrogen is in contact interaction with one of the crown ether oxygen atoms. Since it is attached to a carbon atom having three electron-withdrawing substituents it is sufficiently acidic to form a hydrogen bond.

The most important feature of this structure *viz.* a possible π–π interaction between the naphthalene (π-base) and the ester group (π-acid), explains the higher observed chiral recognition with amino-acid ester salts than with α-phenylethylammonium salts because this fourth interaction will further structure the complexes.

The observed effect of the substituents in the aryl groups, i.e. a higher chiral recognition if the substituent is electron-donating, was accounted for by a reduction of the π-acidity of the ester group and reduction of the proton acidity of the NCH group. The latter would destabilize the *RR*-(L) diastereomer more than the *RR*-(D) complex, resulting in an enhanced chiral recognition for *p*-hydroxyphenylglycine methyl ester salts (Tables 65, 67) by crown ethers [284] and [290]. In earlier papers the increased EDC values with *p*-hydroxyphenylglycine methyl ester salts were attributed to increased π–π repulsion of the aryl and naphthalene rings, thus destabilizing the less favourable diastereomeric complex (Helgeson *et al.*, 1974). The model also accounts for the strikingly different degree of enantiomeric preference of crown ethers [284] and [285] (Peacock *et al.*, 1978). Firstly, the methyl group at one face of crown ether where the complexation occurs limits the available space at one of the two different cavities, and, secondly, the methyl group on the opposite side of the molecule limits the chances for the less stable enantiomeric complex to stabilize itself by widening the upper cavity where the amino-acid ester salt is bound (see Fig. 4). This will result in a larger difference in complex stability between the *SS*-(D) or *RR*-(L) and *SS*-(L) or *RR*-(D) enantiomers. Other effects e.g. the influence of *p*-phenyl substituents hardly affect the enantiomeric differentiation in this series.

In complexes of crown ethers [303] to [305] (types IV and V) both large groups are located in the larger of the two non-equivalent chiral cavities (Kyba *et al.*, 1978). This was concluded from ^{1}H nmr spectra of the diastereomeric complexes of [303] with α-phenylethylammonium salts. In the spectra the position of the methyl protons in both diastereomeric complexes is the same, in contrast to the methyl protons in complexes of [284] or [285] with the same salt. In the latter the upfield shift of the methyl protons in the more stable *SS*-(*R*) enantiomeric complex differs from that in the less stable *SS*-(*S*) enantiomer. Taking into account the available data summarized above, the tentative conclusion seems to be that the simple steric model (see structure

[306]) will predict the more stable diastereomeric complex if the following conditions are fulfilled:

(*a*) Medium and large groups differ substantially in size [high $\Delta(\Delta G°)$ values].

(*b*) Steric barriers extend in space without interfering too much with the complexation process.

(*c*) Conformational mobilities of large and medium-sized groups are limited. Superimposed on the steric model there are electronic effects exerted by L on the acidities of the carboxylate and the NCH groups. These may enhance or lower the degree of enantiomeric preference. Another relatively small factor might be the packing density of the atoms in the complexes (differences between methyl and t-butyl esters).

CHIRAL RECOGNITION BY CROWN ETHERS WITH ASYMMETRIC CARBON ATOMS

Chiral recognition of (*RS*)-α-phenylethylammonium and (*RS*)-phenylglycine methyl ester salts by crown ethers in which chiral carbohydrate structure elements are incorporated has been reported by Stoddart *et al.* (Curtis *et al.*,

TABLE 74

Enantiomeric differentiation between (*RS*)-α-phenylethylamine ($R' = CH_3$) and (*RS*)-phenylglycine methyl ester salts ($R' = COOCH_3$)[a]

Crown ether	R′	X	R:S	$\Delta(\Delta G°)$ /kcal mol^{-1}
LL-[252]	CH_3	PF_6^-	40:60	−0.24
DD-[307]	CH_3	PF_6^-	62:38	−0.30
LL-[308]	$COOCH_3$	ClO_4^-	60:40	−0.24

[a] Curtis *et al.*, 1977a,b

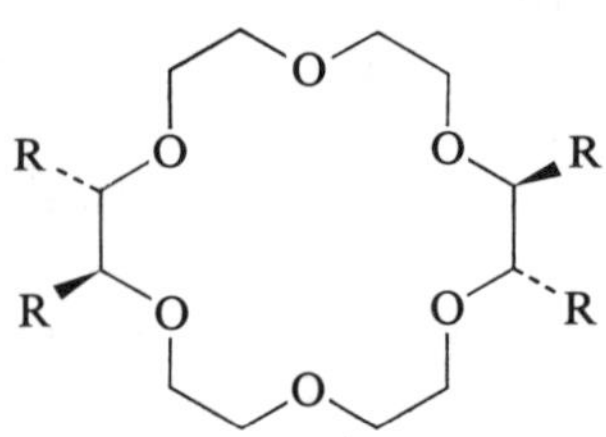

[252], R = $CH_2OCH_2C_6H_5$

[307], R =
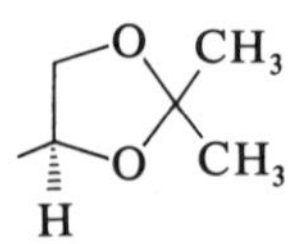

[308], R =
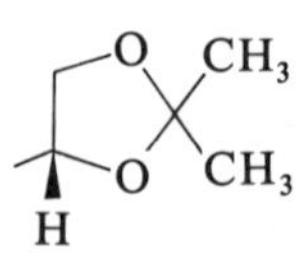

1977a,b). Only in a few cases could enantiomeric differentiation be detected by ^{1}H nmr spectroscopy (Table 74). Considering the above-mentioned results obtained by Cram and his group, this is not surprising because of the low degree of structuring of these diastereomeric complexes.

CHIRAL RECOGNITION OF ANIONS

A different concept of chiral recognition was used by Lehn *et al.* (1978) for the differentiation between pairs of enantiomeric anions. Following the terminology used for metallo-enzymes, the chiral crown ether [309] acts as an apo-receptor, complexing a metal cation and thus becoming a chiral metal receptor that may discriminate between enantiomeric anions (cascade-type complexation). Extraction experiments with racemic mandelic acid dissolved in

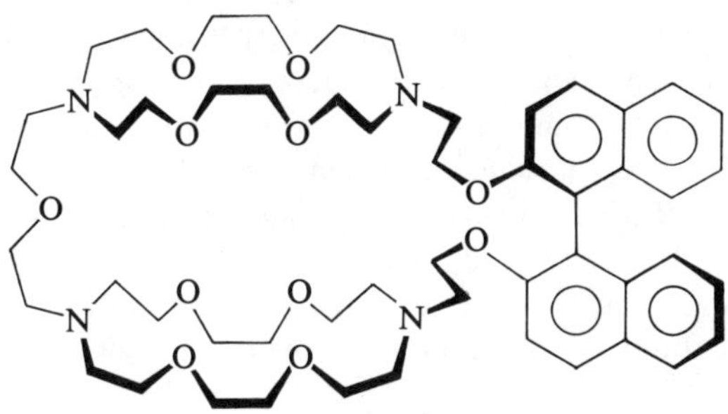

[309]

basic aqueous solutions (Bu_4N^+, Na^+, NH_4^+, K^+, Rb^+, Cs^+) and *S*-[309] in chloroform showed that, except when the bulky Bu_4N^+ is the base cation, the mandelic anion was extracted into the organic phase. Only in the case of potassium and cesium did ^{1}H nmr spectroscopy reveal that chiral discrimination had taken place (Table 75); in the case of K^+ the (−)-mandelate is preferentially extracted, whereas with Cs^+ the (+)-enantiomer dominates.

TABLE 75

Chiral recognition of racemic mandelic acid in the presence of alkali metal cations by [309][a]

Cation	e.e. /%	Configuration of more stable enantiomeric complex
Na^+	15	(D)-*S*
NH_4^+	0	—
Rb^+	0	—
Cs^+	10	(L)-*S*

[a] Lehn *et al.*, 1978

With the larger racemic α-hydroxy-1-naphthaleneacetic acid too, extraction takes place in the presence of [309] and the appropriate cations but enantiomeric differentiation is not observed. These results were confirmed in transport experiments in which the alkali mandelate is carried through a liquid membrane of [309] dissolved in chloroform. Lehn *et al.* (1978) explain these observations in terms of an ion pair included in the cavity of the crown ether. The reversal of chiral recognitions between potassium and cesium mandelate of 25% indicates that the structures of the two complexes are different.

ENZYME-TYPE CATALYSIS BY CROWN ETHERS

This section deals with a type of catalysis by crown ethers that is aimed at the mimicry of enzyme-catalysed reactions. In these reactions the crown ether has the function of bringing substrate and reagent together in close proximity and also of providing the necessary catalytic sites in a trinuclear complex. The work reported so far deals with crown ethers in which the reagent (or substrate) is already built-in (binuclear complex) or with reactions of crown ether/substrate complexes in solutions of reagents. Two types of reaction have been studied in detail, *viz.*, intermolecular hydrogen-transfer in analogy with NAD(P)H, and the acyl-transfer reactions of amino-acid esters analogous to the enzymes trypsin and papain. In the latter reactions not only catalysis of the reaction but also enantiomeric differentiation by the chiral crown ethers is observed.

Intermolecular hydrogen-transfer reactions

The macrocyclic 1,4-dihydropyridine-*bis*-lactone [310] that forms complexes with salts (van der Veen *et al.*, 1978) reacts with phenacylsulfonium salts as shown in (53) to yield the corresponding N-methylpyridinium-*bis*-lactone [311], a carbonyl compound and a sulfide (Van Bergen and Kellogg, 1977). Isomerization of the 1,4-dihydropyridine ring in [310] also takes place, as in other 3,5-disubstituted 1,4-dihydropyridines, to give [312] (Van Bergen and Kellogg, 1976). The crown ether (ester) [310] reacts much faster with sulfonium salts [313a] than 3,5-bis(ethoxycarbonyl)-1,2,6-trimethyl-1,4-dihydropyridine (yields given in parentheses). Extrapolation of the data reported indicates that at 75°C the reaction between [310] and [313a] is a factor of 2.7×10^2 faster. The activation parameters for the reaction of [310] with [313a] ($\Delta H^{\ddagger} = 31.6$ kcal mol^{-1} and $T\Delta S^{\ddagger} = +8.6$ kcal mol^{-1} at 25°C) differ from those of the reaction of [313a] with 3,5-bis(ethoxycarbonyl)-1,2,6-trimethyl-1,4-dihydropyridine ($\Delta H^{\ddagger}_{25°} = 17.7$ kcal mol^{-1} and $T\Delta S^{\ddagger}_{25°} = -8.3$ kcal mol^{-1}). Obviously the reaction of [310] takes place not in a bimolecular reaction but in a preformed binary complex such as the one depicted in [314]. The enhanced rate is largely

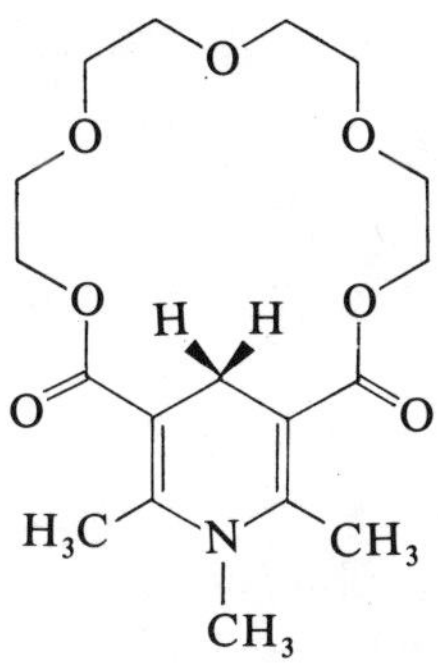

[310]

+

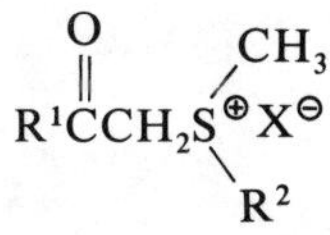

[313]

a. $R^1 = C_6H_5$, $R^2 = C_6H_5$, $X = BF_4$

b. $R^1 = C_6H_5$, $R_2 = CH_3$, $X = ClO_4$

c. $R^1 = CH_3$, $R^2 = C_6H_5$, $X = ClO_4$

d. $R^1 = OCH_3$, $R^2 = C_6H_5$, $X = BF_4$

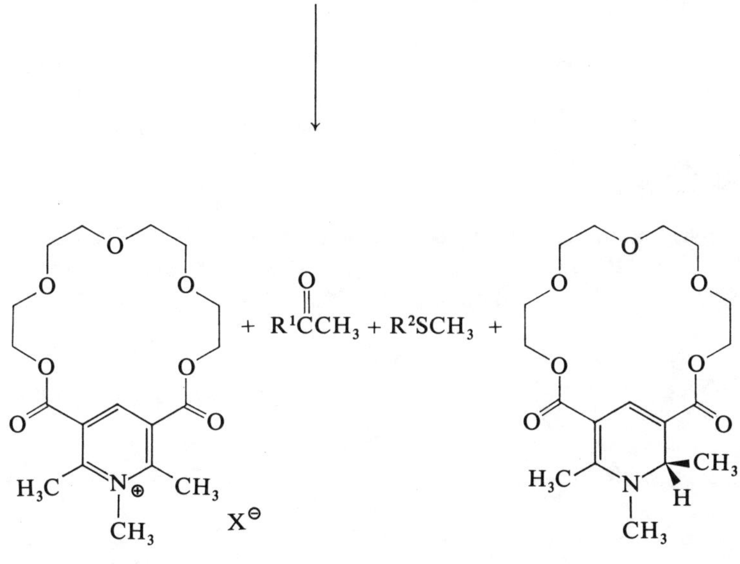

(53)

[311] [312]

[311]		[312]	
43%	(15%)	43%	(86%)
50	(4)	33	(0)
26	(11)	40	(12)
12	(14)	79	(38)

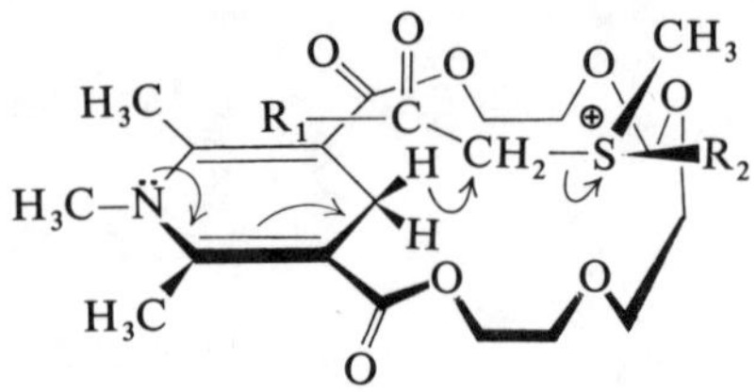

[314]

due to the favourable entropy contribution. Not all sulfonium salts react faster with [310]. Only [313b] reacts with a comparable yield but, [313c] and [313d] undergo reaction with [310] and with the open-chain reference compound at the same rates. Enhanced rates of hydrogen transfer were also reported by Behr and Lehn (1978) in complexes of 18-crown-6 substituted with four side-chains having terminal 1,4-dihydropyridine groups [315] and N-(ω-

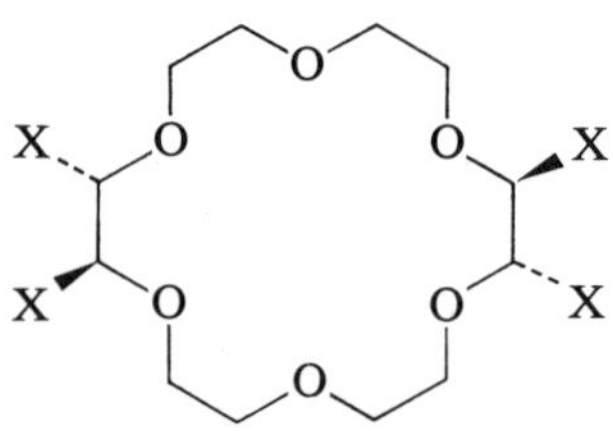

[315], X = $CONH(CH_2)_2N$ 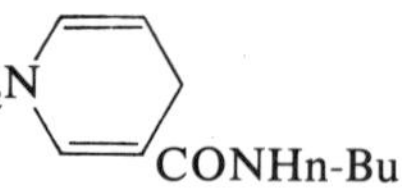

[316], X = $CONH(CH_2)_2$(3-indole)

aminoalkyl)pyridinium salts. Previously Behr *et al.* (1976) observed charge-transfer bands in the UV spectra of complexes of [316] and N-(ω-aminoalkyl)pyridinium salts [317] and concluded that the indole moiety of the crown

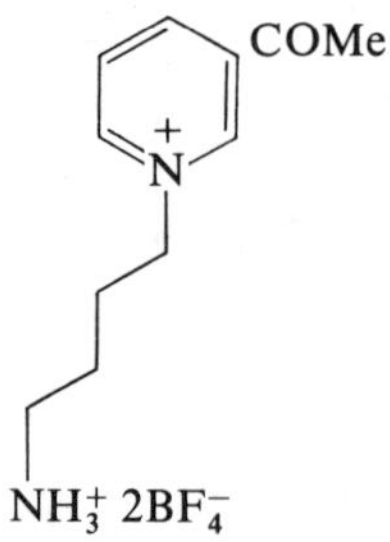

[317]

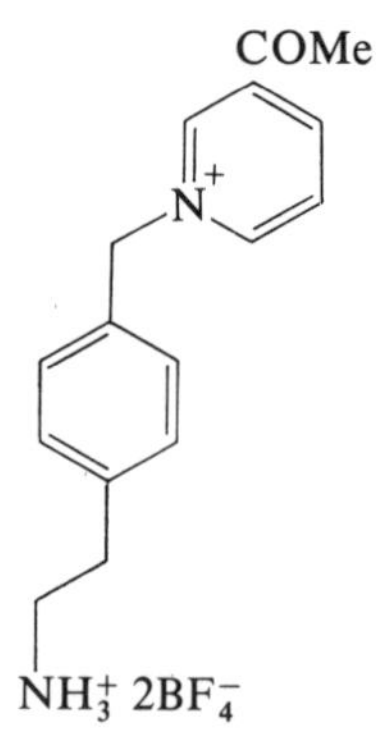

[318]

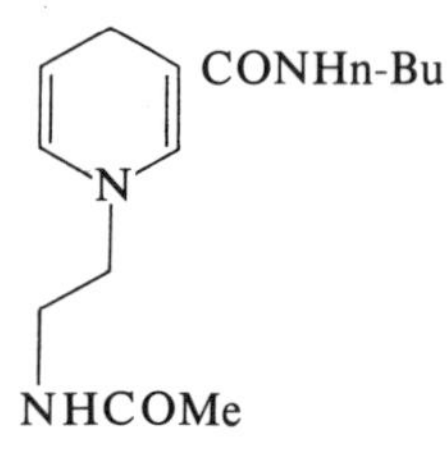

[319]

ether was very close to the pyridinium moiety of the salt. When such pyridinium salts are mixed with solutions of crown ether [315] in acetonitrile, reduction to the corresponding 1,4-dihydropyridines occurs immediately. The kinetics of the reactions are first order, which proves that the transfer of hydrogen takes place in a binuclear complex; $k([317]) = 1.2 \times 10^{-4}\ s^{-1}$ and $k([318]) = 2.5 \times 10^{-4}\ s^{-1}$. Inhibition of the reaction is observed when excess of potassium tetrafluoroborate is added and the transfer of hydrogen from [315] to pyridinium salts then follows second-order kinetics at identical rates for both substrates ($k_2 = 1.2 \times 10^{-2}\ M^{-1}\ s^{-1}$). This second-order rate constant is the same as for hydrogen transfer between [317], [318], and reference compound [319]. The *bi*molecular reaction is inhibited when instead of potassium salts 1,4-diaminobutane salts are added to crown ether [315] or reference compound [319]. Probably, the formation of a positively charged pyridinium group attached to the crown ether ring is disfavoured by electrostatic repulsion by the complexed dication ($k < 10^{-4}\ M^{-1}\ s^{-1}$). The structure of the complex in which the hydrogen transfer is likely to take place is shown in Fig. 5.

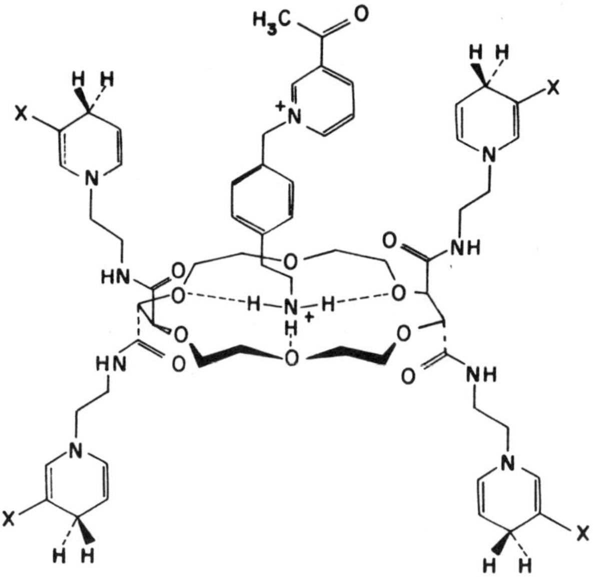

FIG. 5 Schematic representation of the complex of receptor [315]; X = CONHn-Bu with substrate [318]

Catalysis of acyl-transfer reactions

Several crown ethers that possess side chains with terminal mercapto groups enhance the rate of transesterification of amino-acid *p*-nitrophenyl esters. Matsui and Koga (1978) reported the reactions of a number of amino-acid *p*-nitrophenyl ester hydrobromides dissolved in mixtures of ethanol and dichloromethane (1 :4) and buffered with acetic acid and pyridine (pH 4.60 in water) to protonate the amino groups in the presence of 2,3-substituted 18-crown-6 macrocycles.

TABLE 76

Pseudo first-order rate constants (10^5 k/s^{-1}) for *p*-nitrophenol release from amino-acid ester hydrobromides ($RCOOC_6H_4NO_2$)[a,b]

R	Crown ether					
	None	18-crown-6	18-crown-6 + BuSH[c]	[320][d]	[321][d]	[322][d]
$H_3\overset{+}{N}-CH_2-$	3	0.9	1	1170	50	2500
$CH_3\overset{+}{N}H_2-CH_2-$	5	5	4	6	4	37
$H_3\overset{+}{N}-(CH_2)_2-$	<0.1	<0.05	<0.05	0.4	7	2
$H_3\overset{+}{N}-(CH_2)_3-$	310	1	0.9	6	42	41
$H_3\overset{+}{N}-(CH_2)_5-$	<0.05	<0.05	<0.05	<0.05	<0.05	<0.05

[a] Matsui and Koga, 1978
[b] 20% $EtOH-CH_2Cl_2$, buffered with 0.01 M AcOH and 0.005 M pyridine (pH 4.60 in water) at 25°C, 10^{-4} M in substrates, 5×10^{-3} M in crown ethers
[c] 10^{-2} M
[d] Corrected for buffer solvolysis in the presence of 18-crown-6

The results in Table 76 show that the rate of transesterification is very much dependent on the structural relationship between the amino-acid ester and the crown ether. The conclusion that the reactions are fast when they occur in a

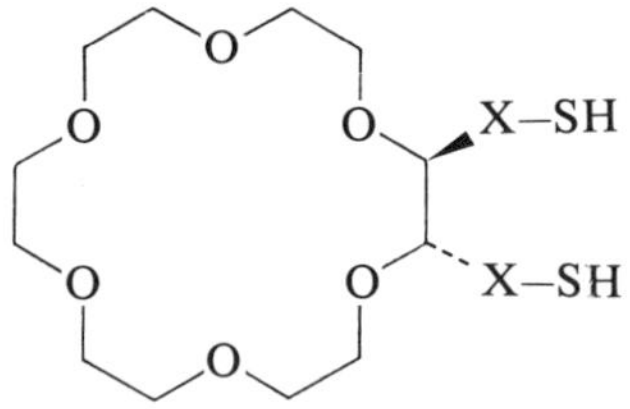

[320], $X = -CH_2-$
[321], $X = -(CH_2)_3-$
[322], $X = -CH_2O(CH_2)_2-$

binary complex is supported by the following observations. The stability constant of crown-ether complexes with dialkylammonium salts are small compared with those of complexes with alkylammonium salts. This is reflected in the rate factors of 66 and 195 between N-methylglycine and glycine *p*-nitrophenyl esters with the more effective catalysts [322] and [320]. In addition, the length of the crown-ether substituent and the length of the amino-acid ester chain, together determining the positions of the ester and catalytic mercapto groups relative to each other, influence the rate. The striking differences in rate observed between reactions with [321] and [322] are attributed to a larger conformational freedom of the side chains and to a possible additional co-ordination of the amino-acid ester by the side-chain oxygen atom. A similar enhanced structuring of a crown ether/amino-acid complex by a side-chain oxygen atom was noted by Timko *et al.* (1978) in chiral recognition experiments (*vide infra*).

Chiral discrimination between enantiomeric amino-acid *p*-nitrophenyl ester hydrobromides in addition to enhanced rate of transacylation were reported by Chao and Cram (1976) for chiral 3,3′-*bis*(mercaptomethyl)dinaphthyl-20-crown-6 [323]. Compared with a non-cyclic reference compound (*S*)-[324] the rates for a series of amino-acid esters are enhanced by factors of 10^2 to 10^3, except for L-proline. This once more demonstrates that reaction takes place in

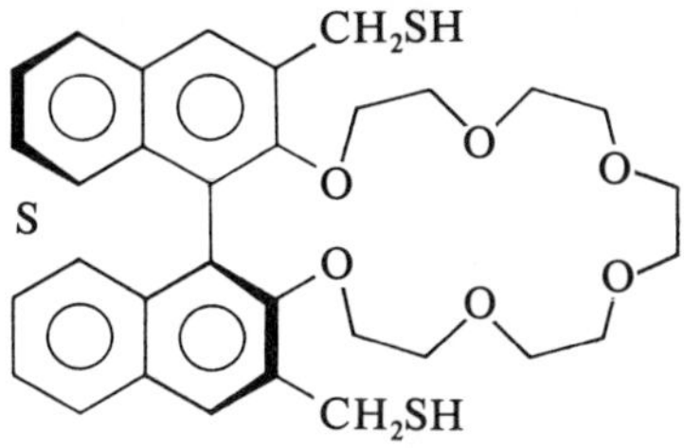

(*S*)-[323]

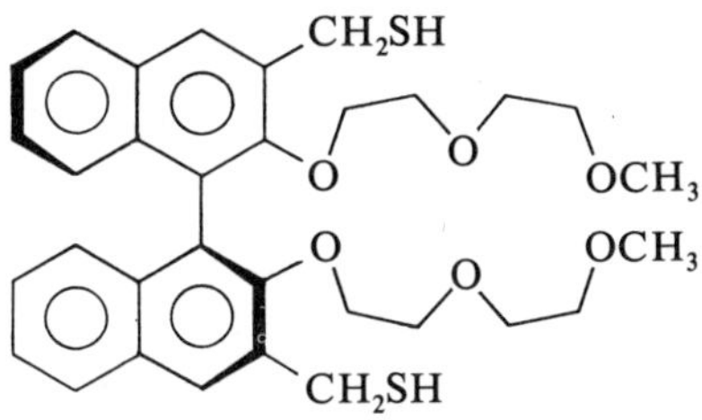

(*S*)-[324]

a binary complex of crown ether and amino-acid salt (Table 77). Thioesters are formed as intermediates by reaction of the complexed ester with the crown ether mercapto groups. The polarity of the medium in which the transacylation takes place plays an important role. The rate factors defined as the rate in the presence of [323] divided by the rate in the presence of [324] are much lower in the more hydrophilic medium (C) than in the more lipophilic medium (A). This is due to the fact that the complexes in the more hydrophilic media are less structured because of hydrogen-bond formation between the ammonium group and solvent molecules.

The rates of reaction of both enantiomers of amino-acid esters in the presence of (*S*)-[324] are the same, but with (*S*)-[323] they are in most cases different. The reactions of L-amino acid esters in the presence of (*S*)-[323] are faster than those in the presence of (*R*)-[323] by factors of 9.2 (R = i-Pr), 8.2 (R = $C_6H_5CH_2$) and 6.0 (R = i-Bu). No difference in rates is observed for L-alanine *p*-nitrophenyl ester. The results were explained in terms of the formation of diastereomeric tetrahedral intermediates [325] and [326]. The bulk of the group R will determine how much the complex stability of the (D)-complex decreases relative to that of the (L)-complex, which difference is reflected in the activation energy of the rate-determining step.

TABLE 77

Rate comparisons for liberation of *p*-nitrophenol from 10^{-4} M solutions of amino-acid ester salts in the presence of 5.0×10^{-3} M [323] or [324] at 25.0°C[a]

Medium[b]	Polyether present	$RC^*H(NH_3^+)CO_2Ar$ R	Config.	$10^3k/s^{-1}$	Rate factors
A	None	H	—	2.6	1
A	[323]	H	—	>702	>270
A	[324]	H	—	5.5	2.1
A	18-crown-6 [3]	H	—	8.1	3.1
A	(*S*)-[323]	Prolyl	L	16	0.8
A	(*S*)-[324]	Prolyl	L	21	
A	(*S*)-[323]	H	L	>700	>130
A	(*S*)-[324]	H	L	5.4	
A	(*S*)-[323]	CH_3	L	≳700	>130
A	(*S*)-[324]	CH_3	L	5.4	
A	(*S*)-[323]	$(CH_3)_2CH$	L	13	160
A	(*S*)-[324]	$(CH_3)_2CH$	L	0.08	
A	(*S*)-[323]	$C_6H_5CH_2$	L	200	500
A	(*S*)-[324]	$C_6H_5CH_2$	L	0.41	
A	(*S*)-[323]	$(CH_3)_2CHCH_2$	L	≳700	≳1100
A	(*S*)-[324]	$(CH_3)_2CHCH_2$	L	0.6	
A	(*S*)-[323]	$C_6H_5CH_2$	D	25	69
A	(*S*)-[324]	$C_6H_5CH_2$	D	0.36	
B	(*S*)-[323]	CH_3	L	70	1
B	(*R*)-[323]	CH_3	L	70	
B	(*S*)-[323]	$(CH_3)_2CHCH_2$	L	~70	~6
B	(*R*)-[323]	$(CH_3)_2CHCH_2$	L	11	
B	(*S*)-[323]	$C_6H_5CH_2$	L	340	8.3
B	(*R*)-[323]	$C_6H_5CH_2$	L	41	
B	(*R*)-[323]	$C_6H_5CH_2$	D	340	8.1
B	(*S*)-[323]	$C_6H_5CH_2$	D	42	
B	(*S*)-[323]	$(CH_3)_2CH$	L	22	9.2
B	(*R*)-[323]	$(CH_3)_2CH$	L	2.4	
C	(*S*)-[323]	H	—	110	20
C	(*S*)-[324]	H	—	5.4	
C	(*S*)-[323]	$C_6H_5CH_2$	L	18	8
C	(*S*)-[324]	$C_6H_5CH_2$	L	2.2	

[a] Chao and Cram, 1976

[b] A 20% EtOH in CH_2Cl_2 (by vol.) buffered with 0.2 M AcOH and 0.1 M $(CH_3)_4NOAc$; B 20% EtOH in CH_2Cl_2 (by vol.) buffered with 0.3 M AcOH and 0.1 M $(CH_3)_4NOAc$; C 40% H_2O in CH_3CN (by vol.) buffered with 0.2 M AcOH and 0.2 M NaOAc. These buffers in water give pH 4.8

Crown ethers [327] in which (L)-cysteinyl residues are connected with the polyether ring via amide bonds in the 2, 3, 11, and 12-positions exhibit

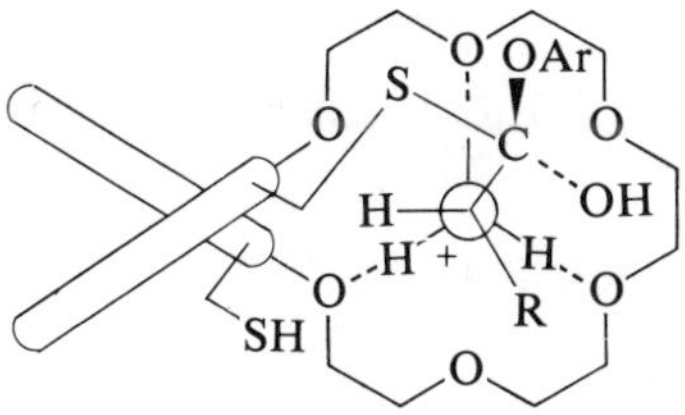

[325], (L)-complex

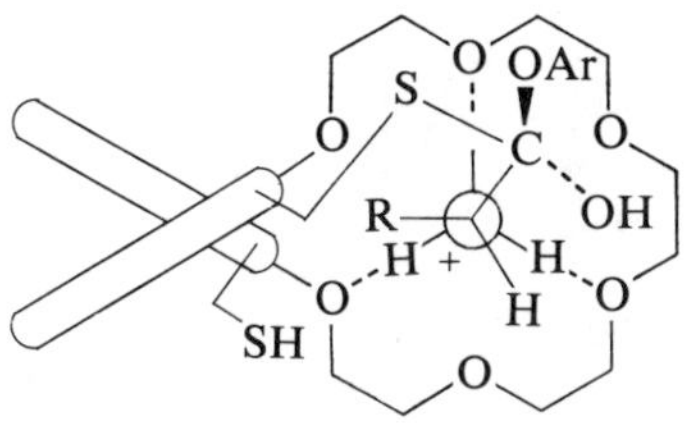

[326], (D)-complex

enhanced rates of thiolysis and high degrees of enantiomeric differentiation of amino acid and dipeptide *p*-nitrophenyl esters (Table 78; Lehn and Sirlin, 1978). Similar observations to those described by Cram and Chao such as inhibition of the reaction by potassium salts, low rates of reaction of (L)-*prolyl*glycine *p*-nitrophenyl ester and no enhanced rates when the thiol groups in [327] are benzylated underline the postulated mechanism of the reaction taking place in a binary complex. Interestingly, reaction of [327] with N-carbobenzoxyglycine *p*-nitrophenyl ester, which will not form a complex, is accelerated by a factor of 15 when potassium bromide is added. Obviously complexation of a cation enhances the rate of the "*inter*molecular" reaction, probably by lowering the p*K* of the mercapto groups. Consequently, complexation of an ammonium salt will have a similar effect and therefore the enhanced rates of thiolysis in the complexes may not entirely be due to stabilization of the tetrahedral intermediates such as [325] and [326].

Lehn and Sirlin found a very high degree of enantiomeric differentiation in the reaction of [327] with racemic dipeptide *p*-nitrophenyl esters [Gly-(DL)-Phe-OPNP]. Gly-(L)-Phe-OPNP reacts at approximately the same rate as the achiral Gly-Gly-OPNP, but the D-enantiomeric dipeptide ester is converted at a

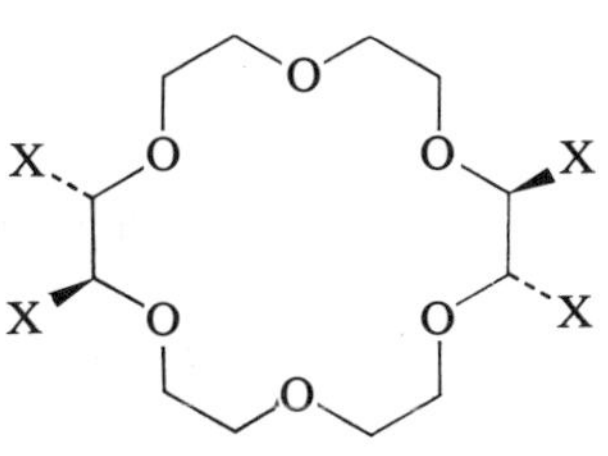

[327], X = —CONH—(L)—CH(CH_2SH)—CO_2Me

TABLE 78

Kinetic data for release of *p*-nitrophenol from amino acid and dipeptide ester salt substrates in the presence of the macrocyclic reagent [327][a]

Run[b]	Substrate[c] R—OPNP, HBr	Rate constant, $10^5\ k_{\psi}/s^{-1}$ [d] Solv.	[327]	([327] + K^+)	Relative rates[e]
1A	Gly	4.8	400	27	15
2A	L- or D-Phe	2.9	19	34	0.6
3A	β-Ala	7.8	18	27	0.7
4A	Gaba	75	52	—	—
5A	Gly-Gly	4.8	230	2.9	80
6A	Gly-(L)-Phe	2.4	150	1.5	100
7A	Gly-(D)-Phe	2.4	23	0.7	35
5B	Gly-Gly	390	235 000	1700	140
					5B/9B = 1900
8B	Gly-β-Ala	4.3	120	38	3
9B	(L)-Pro-Gly	115	125	590	0.2
10B	CbO-Gly	5.5	14.5	220	0.07
5C	Gly-Gly	59	2600	—	5C/9C = 15 000
6C	Gly-(L)-Phe	74	1600	—	6C/7C = 50
7C	Gly-(D)-Phe	74	30	—	—
9C	(L)-Pro-Gly	77	0.17	—	—
6D	Gly-(L)-Phe	15	1500	—	—
7D	Gly-(D)-Phe	15	16.5	—	—

[a] Lehn and Sirlin, 1978
[b] The number characterizes the substrate; capital letter indicates the medium. A: $MeOH—CH_2Cl_2—H_2O$ 78.5 : 20 : 1.5, Py—PyHBr buffer 0.05 M, pH = 6.1; B: MeOH—dimethylformamide—H_2O 78.5 : 20 : 1.5, AcOH—$AcONMe_4$ buffer 0.02 M, pH = 4.8; C: CH_2Cl_2—MeOH—H_2O 97.9 : 2 : 0.1, CF_2CO_2H—N—ethylmorpholine buffer 0.3 M, pH = 7.0; D: CH_2Cl_2—EtOH 95 : 5, CF_2CO_2H—N—ethylmorpholine buffer 0.03 M, pH = 7.0; the pH indicated corresponds to that of an aqueous solution containing the same buffer concentration as used here. In media A and B, substrate: 10^{-4} M, [327]: 3.5×10^{-3} M, KBr: 1.4×10^{-2} M; in media C and D, substrate: 5×10^{-5} M, [327] and [327], $SCH_2C_6H_5$ replaces SH: 3.5×10^{-4} M.
[c] All substrates are in the form of ammonium hydrobromides. Gly: glycyl, $H_2N—CH_2—CO—$; Phe: phenylalanyl, $H_2N—CH(CH_2Ph)—CO—$; β-Ala: β-alanyl, $H_2N—CH_2—CH_2—CO—$; Gaba: γ-aminobutyryl. $H_2N—[CH_2]_3—CO—$; Pro: prolyl, $HN—[CH_2]_3—CH—CO—$; CbO: benzyloxycarbonyl $PhCH_2OCO—$
[d] Pseudo-first order rate constants at 20°C for release of *p*-nitrophenol
[e] Relative rates refer to the ratio of [327]/[327 + K^+]

rate that is 50–90 times lower. By using this difference in rate a kinetic resolution of the dipeptide esters could be achieved.

The chiral discrimination may be caused by differences in complexation between the enantiomeric substrates in the ground state or in reactivity between the complexes.

Miscellaneous

A catalytic effect that may be of the same type as the mimicry of acyl transferase enzymes is reported by Gandour *et al.* (1978). They describe the rate enhancement of the aminolysis of *p*-nitrophenyl acetate in the presence of bis(2,6-pyridinyl) crown ethers [328] and [329] in chlorobenzene. The rate of

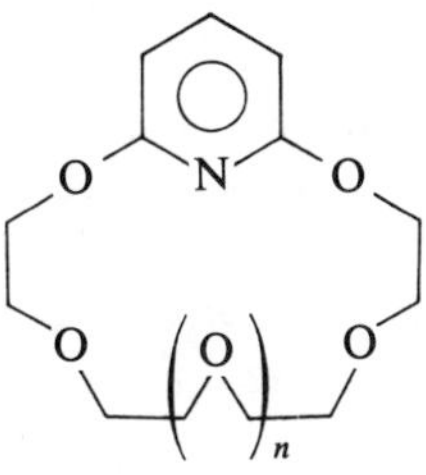

[328]

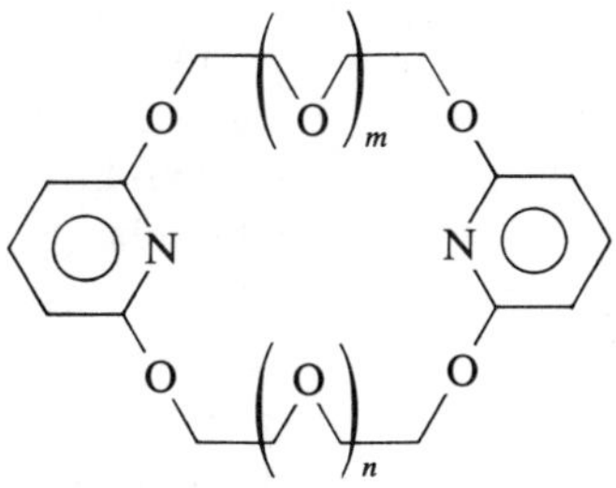

[329]

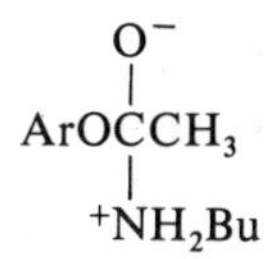

[330]

the catalysed reaction increases with increasing ring size of the macrocyclic ligand (Table 79). The catalysis is ascribed to stabilization of a polar intermediate [330] by inclusion into the polar crown-ether cavity without specific co-ordination. The results in the table indicate that the pyridine nitrogen atoms are not necessary for the catalysis.

TABLE 79

Catalytic rate constants for butylaminolysis of *p*-nitrophenyl acetate in chlorobenzene at 25 °C[a]

Base	Ring size	$10^3k'$ /$M^{-2}\,s^{-1}$	Rel. rate	$10^3k'$/oxygen atoms
[328], $n = 1$	16	7.2	1	1.4
[328], $n = 2$	19	27	3.8	4.5
[328], $n = 3$	22	62	8.6	8.9
[329], $n = 2$, $m = 3$	29	160	22	17.7
[329], $n = m = 3$	32	210	29	21.0
[329], $n = m = 4$	38	290	40	24.2
[329], $n = m = 5$	44	430	60	30.7
Dibenzo-18-crown-6 [11]	18	34	—	5.6
2,6-Dimethoxypyridine	—	No catalysis	—	—
Butylamine	—	60	—	—
2,6-Dimethylpyridine	—	9	—	—
Tetrahydrofuran	—	4	—	4
H_2O	—	290[b]	—	290

[a] Gandour *et al.*, 1978

[b] Derived by dividing the spontaneous rate constant by $[H_2O] = 55.5$ M

Designed to be a model for metallo-proteins Chang (1977) has synthesized a so-called "crowned" porphyrin [331] that may have the ability to bind a transition-metal ion and an alkali or alkaline-earth cation simultaneously. He found that, in the presence of 1-methylimidazole or 1-isobutylimidazole, Fe(II) complexes are formed that bind oxygen at room temperature. These complexes are more stable than simple hemes because under the same conditions they are totally oxidized within 30 seconds. The oxygenated complex of [331] with Fe(II) in the presence of 1-triphenylmethylimidazole is stable at 25°C ($t_{\frac{1}{2}} > 60$ minutes). This indicates that the nitrogen co-ordinates on the free side of [331] with the oxygen molecule encapsulated in this cavity. This protects the oxygenated complex from bimolecular oxidation.

Related to these "enzyme"-catalysed reactions are electroreductions of acetophenone in the presence of chiral crown ethers. The low optical yields (<3%) are attributed to association of the prochiral substrate and the chiral crown ether salt complex in the electrochemical double layer (Horner and Brich, 1978).

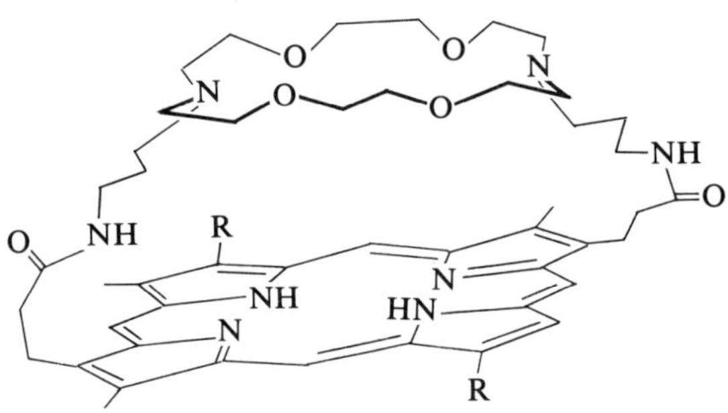

[331], R = n-hexyl

7 Complexes with arenediazonium salts

Cram *et al.* (Gokel and Cram, 1973; Kyba *et al.*, 1977b) found that several crown ethers *viz.* 18-crown-6, dinaphthyl-20-crown-6 and dinaphthyl-23-crown-7 solubilize 4-substituted and 3,4-disubstituted aryldiazonium tetrafluoroborates in chloroform. The ratios of crown ether to diazonium salt vary with the crown-ether structure and with the electron-donating or -withdrawing nature of the substituent(s) from 0.3 to 1.0. However, 2,6-dimethylbenzenediazonium tetrafluoroborate is not complexed by crown ethers because the methyl groups would cause steric hindrance in a complex such as [332]. Neither does the *bis*(dinaphthyl)-22-crown-6 form a complex in chloroform, which is most probably due to the lower basicity of the crown-ether aryl oxygen atoms. Complexes of arenediazonium salts with dinaphthyl-20-crown-6 are yellow or red whereas 18-crown-6 forms colourless complexes. The authors suggest that the dinaphthyl group acts as a π-base and the

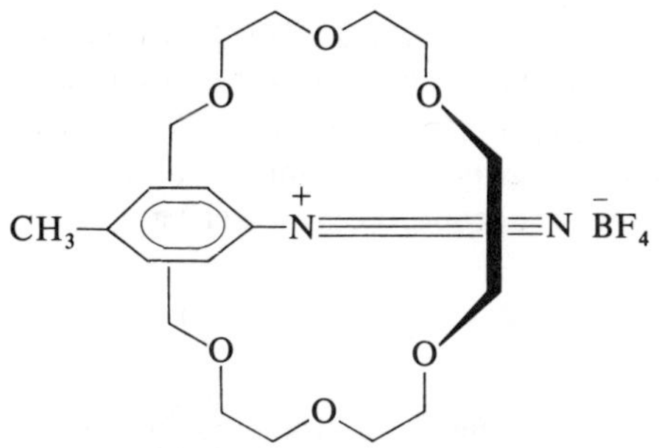

[332]

arenediazonium salts act as a π-acid, which results in a charge-transfer interaction that may provide additional stabilization for the complex. Small changes are observed in the ^1H nmr spectra upon complexation of the crown ethers but more relevant information about the complex structure has been obtained from IR spectroscopy in the solid state or in solution. The stretching frequency of the N≡N bond in the complexed diazonium salt is shifted to *higher* values. (Usually complexation of a diazonium salt results in lowering of this frequency.) The respective values for complexed and non-complexed salt are 2317 cm^{-1} and 2285 cm^{-1} for the dicyclohexyl-18-crown-6 complex with benzenediazonium hexafluorophosphate (Haymore *et al.*, 1975), 2307 cm^{-1} and 2269 cm^{-1} for 18-crown-6 complexed with 4-t-butylbenzenediazonium tetrafluoroborate in CH_2Cl_2 solution (Korzeniowski *et al.*, 1977), and 2306 cm^{-1} and 2277 cm^{-1} for the same complex in a nujol mull (Bartsch *et al.*, 1976). Complexation of e.g. 4-t-butylbenzenediazonium tetrafluoroborate and 18-crown-6 can also be detected by ^{13}C nmr spectroscopy. The most pronounced chemical shift differences ($\Delta\delta \sim 3$ ppm) are found for the C–1, C–2 and C–4 carbon atoms (Korzeniowski *et al.*, 1977).

So far only a few quantitative data on the thermodynamic stability of arenediazonium salts and crown ethers have been reported. Bartsch *et al.* (1976) calculated the value of the association constant of the complex of 18-crown-6 and 4-t-butylbenzenediazonium tetrafluoroborate from kinetic data on the thermal decomposition of the complex, $K_a = 1.56 \times 10^5$ l mol^{-1} in 1,2-dichloroethane at 50°C. Compared with the corresponding linear polyether this is at least a factor of 30 higher (Bartsch and Juri, 1979).

Izatt *et al.* (1978c) have determined the values of log K, ΔH, and $T\Delta S$ for diazonium complexes of 18-crown-6 in methanol at 25°C. These values were 2.26, −10.03 kcal mol^{-1}, and −6.95 kcal mol^{-1} for 4-methylbenzenediazonium, and 2.37, −14.0 kcal mol^{-1}, and −10.8 kcal mol^{-1} for benzenediazonium. Strongly electron-donating groups such as Et_2N depressed complexation to such an extent that log K values could not be determined. The low stability of the benzenediazonium as compared with the anilinium complex of 18-crown-6 (log K, 3.80; ΔH, −9.54 kcal mol^{-1}; $T\Delta S$, −4.36 kcal mol^{-1}) is entirely due to the very unfavourable entropy contribution. In agreement with this result, Kyba *et al.* (1977b) found a dramatic increase of the solubilization of 4-chlorobenzenediazonium tetrafluoroborate and Korzeniowski *et al.* (1977) an increased degree of complexation in mixtures of 18-crown-6 and 4-t-butylbenzenediazonium tetrafluoroborate by lowering the temperature.

REACTIVITY OF CROWN ETHER COMPLEXES WITH ARENEDIAZONIUM SALTS

The thermal decomposition of arenediazonium tetrafluoroborates is slowed down when the salt is complexed by 18-crown-6 (Bartsch *et al.*, 1976). The kinetic data obtained for the 4-t-butylbenzenediazonium salt at 50°C in 1,2-dichloroethane revealed that the rate of complexed to uncomplexed salt is more than 100. Other crown ethers such as dibenzo-18-crown-6 and dicyclohexyl-18-crown-6 exhibited the same effect but smaller molecules such as 15-crown-5 did not influence the rate at all. It is not only the rate of the Schiemann reaction that is affected by the crown ether; nucleophilic aromatic substitutions by halide ions (Cl^-, Br^-) at the 4-positions in arenediazonium salts are retarded or even entirely inhibited when 18-crown-6 is added. This is attributed to the attenuation of the positive charge at the diazonio group in the complex (Gokel *et al.*, 1977).

On the other hand, Hartman and Biffar (1977) reported that decomposition of arenediazonium tetrafluoroborates in dichloromethane in the presence of copper metal is catalysed by dicyclohexyl-18-crown-6. Electron-withdrawing substituents in the aryl ring enhance the rate of the reaction. The main function of the crown ether is probably to solubilize the salt. The effect of the complexation on the rate was not investigated in detail. Similar enhanced solubilization of diazonium salts in apolar solvents was reported and used by Martin and Bloch (1971) in pyrolysis experiments aimed at the generation of the dehydrocyclopentadienyl anion.

Juri and Bartsch (1979) have studied the coupling of 4-t-butylbenzenediazonium tetrafluoroborate with N,N-dimethylaniline in 1,2-dichloroethane solution. The addition of one equivalent (based on diazonium salt) of 18-crown-6 caused the rate constant to drop by a factor of 10, indicating that complexed diazonium is less reactive than the free cation. Benzenediazonium tetrafluoroborate complexes of crown ethers are photochemically more stable than the free salt. The decomposition into fluorobenzene and boron trifluoride is strongly inhibited but no explanation has been given (Bartsch *et al.*, 1977).

8 Complexes with neutral molecules

A number of complexes of crown ethers with neutral molecules have been reported. Some of them have been isolated and their structure has been resolved by single-crystal X-ray analysis. It seems that the forces by which these complexes are held together are weak hydrogen bonds between guest and host.

Pedersen (1971) reported on a number of crystalline complexes with thiourea and similar molecules and found that the solubility of dibenzo-18-crown-6 [11] in methanol is enhanced by urea and thiourea. The latter observation precludes the possibility of the solid complexes being urea or thiourea inclusion compounds. Factors that are of importance in the complex formation are ring size (dibenzo-12-crown-4 did not form complexes), and steric hindrance in the complex caused by the guest molecule. Thiourea, N-phenylthiourea, and 2-thiazolidinethione form complexes but N,N,N′-trimethylurea, thiocarbanilide and N-methylthiazolidinethione do not.

Similar complexes with non-cyclic polyethers that could not be obtained by Pedersen were recently described by Rasshofer and Vögtle (1978) and Suh and Saenger (1978).

The structure of a 1:2 complex of 18-crown-6 and benzenesulphonamide has been determined by x-ray (Knöchel *et al.*, 1978). It shows that the sulphonamide group is hydrogen-bonded to the crown-ether oxygen atoms.

TABLE 80

Stability constants for complexes of sulfonamides with 18-crown-6 in benzene and in chloroform at 10°C[a]

Sulfonamides	Stability constant K/M^{-1} In benzene	Stability constant K/M^{-1} In chloroform
Sulfamethomidine	34.30	20.52
Sulfamonomethoxine	166.0	14.11
Sulfamethoxazole	—	10.00
Sulfaphenazole	41.95	6.68
Sulfamethoxypyridazine	29.71	10.60
Sulfadimethoxine	50.38	10.52
Sulfamethizole	[b]	11.21
Sulfisoxazole	80.71	13.35
Sulfisomidine	[b]	5.44
Sulfamerazine	[b]	22.38
Sulfathiazole	[b]	21.08
Sulfanilamide	[b]	48.54

[a] Takayama *et al.*, 1977

[b] Could not be determined with accuracy because of the poor solubility of sulfonamides in benzene

The N–H . . . . O interactions are unequal, with H . . . . O distances of 1.994 and 2.337 Å. Complexes of sulphonamides with crown ethers are described by Takayama *et al.* (1977). They investigated the complexation of a number of commercially available drugs (Table 80) with 18-crown-6 in chloroform and benzene. The stoichiometry of these complexes is 1 : 1 and the stability constants vary from 5 to 166. The stability of the complexes is higher in the less polar benzene than in the more polar chloroform. The nature of the binding was not elucidated but it was suggested that complexation occurs via the 4-arylamino group rather than via the sulphonamido group.

Water is solubilized in chloroform by *m*-xylene-crown ethers [256], $R_1 = R_2 = H$ (de Jong *et al.*, 1976a). The results in Table 81 show that complexation increases with the ring size of the crown ether and hence with the number of oxygen atoms in the ring. There is no optimal ring size for complexation of

TABLE 81

Complexation of water by crown ethers in $CDCl_3$ at $22 \pm 1°C$[a]

Crown ether	K_w[b]	R[c]	Crown ether	K_w[b]	R[c]
[256], $n = 2$	8	0.27	$n = 6$	23	0.51
$n = 3$	14	0.38	$n = 7$	47	0.68
$n = 4$	14	0.39	$[Ph(CH_2OCH_2)_3]_2$	7	0.24
$n = 5$	20	0.47	18-crown-6	30	0.60

[a] De Jong *et al.*, 1976a
[b] Association constant (M^{-1}) for the binding of water by crown ethers to form a 1 : 1 complex. Estimated accuracy: 10%
[c] Molar ratio solubilized water/crown ether

water as there is for ammonium salts. A solid complex of the 2,2′-dihydroxydinaphthyl crown ether [57] with one molecule of water was obtained by Koenig *et al.* (1976) and its structure elucidated by Goldberg (1978). Compared with the size of the crown-ether cavity the water molecule is small and therefore only some of the crown-ether oxygen atoms can take part in the binding. The water molecule forms two hydrogen bonds with two crown-ether oxygen atoms; additional stabilization of the complex is provided by a third hydrogen bond between one of the two phenolic hydroxyl groups and the oxygen atom of water. Co-complexation of water with salts and crown ethers is often observed in solution (de Jong *et al.*, 1976a, Iwachido *et al.*, 1976) and in the solid state (Hughes and Wingfield, 1977; Cradwick and Poonia, 1977).

(*R,R*)-*Bis*(dinaphthyl)-22-crown-6 at low temperature (−78°C) in carbon disulfide solution selectively extracts one molecule of deuteromethanol out of a mixture of CH_3OD/D_2O (4/1). The complex may have a structure in which the methyl group is hydrogen-bonded to alternate oxygen atoms of the crown ether ring (Cram *et al.*, 1978a). Complexation of crown ethers with neutral C—H acids, probably via hydrogen bond formation, has been reported for nitromethane (McLachlan, 1974), acetonitrile (McLachlan, 1974; Gokel *et al.*, 1974), malonodinitrile (Kaufmann *et al.*, 1977) and acetic anhydride (McLachlan, 1974).

For two complexes the structures have been elucidated by single-crystal x-ray analysis *viz.*, the 2:1 complex of malonodinitrile with 18-crown-6, which revealed the binding contribution of two non-equivalent hydrogen-bonds with H—O distances of 2.235 and 2.694 Å, and the 2:1 complex of nitromethane and 18-crown-6 of which only the D_{3d} geometry was reported. The acetonitrile complex has been used for the purification of crude 18-crown-6 (Gokel *et al.*, 1974). Dimethyl acetylenedicarboxylate forms complexes with furano-18-crown-6 [208], 18-crown-5 [207], and with 18-crown-6 [3], each time with a 1:1 stoichiometry of the compounds (Timko and Cram, 1974; Timko *et al.*, 1977). The x-ray structure determination (Goldberg, 1975) revealed that in the complex with 18-crown-6 the acetylene is bound to two polyether rings. Only two of the three protons of the methyl ester functions interact with the crown-ether oxygen atoms.

The chemical reactivity of crown-ether complexes with neutral molecules has received little attention. Nakabayashi *et al.* (1976) have reported crown-ether catalysis in the reaction of thiols with 1-chloro-2,4-dinitrobenzene. The catalytic activity was attributed to deprotonation of thiols by dicyclohexyl-18-crown-6 in acetonitrile solution. Blackmer *et al.* (1978) found that the rate of aquation of the cobalt(III) complex [333] increases on addition of

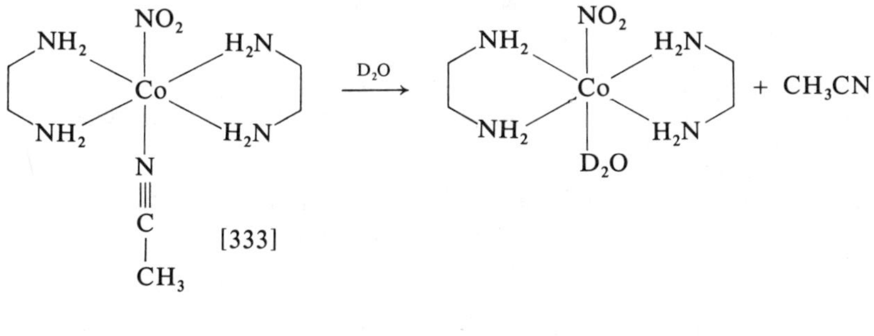

[333]

18-crown-6. The kinetics of the process were consistent with a mechanism in which 18-crown-6 forms a complex with the cobalt-bound acetonitrile, followed by a rate-limiting dissociation of the Co—$NCCH_3$.18-crown-6 linkage. An upper limit of 0.1 M^{-1} was estimated for the binding constant between 18-crown-6 and [333].

Note added in proof

With regard to the chiral recognition by crown ethers D. J. Cram kindly informed us that the EDC value of 38 (footnote *b*, Table 67) proved to be in error, and that the reported *RR-S* configuration in Table 68, footnote *d* and page 403, is still uncertain. Recent work (Peacock *et al.*, 1980) has shown that the chiral recognition of amino acids (page 397 and Table 69) is comparable to that of amino-acid esters. The peculiar optimum in EDC values as a function of acetonitrile concentration (page 401 and Table 72) could not be duplicated.

References

Abraham, M. H., Danil de Namor, A. F. and Lee, W. H. (1977). *J. C. S. Chem. Comm.* 893

Agostiano, A., Caselli, M. and Della Monica, M. (1976). *J. Electroanal. Chem.* **74**, 95

Akabori, S. and Ohtomi, M. (1975). *Bull. Chem. Soc. Jap.* **48**, 2991

Akabori, S., Ohtomi, M. and Arai, K. (1976). *Bull. Chem. Soc. Jap.* **49**, 746

Akabori, S. and Tuji, H. (1978). *Bull. Chem. Soc. Jap.* **51**, 1197

Akabori, S. and Yoshii, T. (1978). *Tetrahedron Lett.* 4523

Akabori, S., Miyamoto, S. and Tanabe, H. (1978). *J. Pol. Sci. Polym. Lett. Ed.* **16**, 533

Akhmetova, N. E., Vlasov, V. M. and Yakobson, G. G. (1978). *Izv. Akad. Nauk USSR, Ser Khim.* 949

Alberts, A. H. and Cram, D. J. (1976). *J. C. S. Chem. Comm.* 958

Alberts, A. H. and Cram, D. J. (1977). *J. Amer. Chem. Soc.* **99**, 3880

Alunni, S., Baciocchi, E. and Perucci, P. (1976). *J. Org. Chem.* **41**, 2636

Alunni, S., Baciocchi, E. and Perucci, P. (1977). *J. Org. Chem.* **42**, 2170

Anderegg, G. (1975). *Helv. Chim. Acta* **58**, 1218

Ando, N., Yamamota, Y., Oda, J. and Inouye, Y. (1978). *Synthesis* 688

Arnaud-Neu, F., Spiess, B. and Schwing-Weil, M.-J. (1977). *Helv. Chim. Acta* **60**, 2633

Arnett, E. M. and Moriarity, T. C. (1971). *J. Amer. Chem. Soc.* **93**, 4908

Arnett, E. M., Chawla, B. and Hornung, N. J. (1977). *J. Solution Chem.* **6**, 781

Baciocchi, E. and Sebastiani, G. V. (1979). *J. Org. Chem.* **44**, 28

Bartsch, R. A. and Wiegers, K. E. (1972). *Tetrahedron Lett.* 3819

Bartsch, R. A. (1973). *J. Org. Chem.* **38**, 846

Bartsch, R. A., Pruss, G. M., Cook, D. M., Buswell, R. L., Bushaw, B. A. and Wiegers, K. E. (1973). *J. Amer. Chem. Soc.* **95**, 6745

Bartsch, R. A. and Shelly, T. A. (1973). *J. Org. Chem.* **38**, 2911

Bartsch, R. A., Mintz, E. A. and Parlman, R. M. (1974). *J. Amer. Chem. Soc.* **96**, 4249

Bartsch, R. A. and Kayser, R. H. (1974). *J. Amer. Chem. Soc.* **96**, 4346

Bartsch, R. A. (1975). *Accounts Chem. Res.* **8**, 239

Bartsch, R. A., Chen, H., Haddock, N. F. and Juri, P. N. (1976). *J. Amer. Chem. Soc.* **98**, 6753

Bartsch, R. A., Allaway, J. R. and Lee, J. G. (1977). *Tetrahedron Lett.* 779

Bartsch, R. A. and Roberts, D. K. (1977). *Tetrahedron Lett.* 321

Bartsch, R. A., Haddock, N. F. and McCann, D. W. (1977). *Tetrahedron Lett.* 3779

Bartsch, R. A. and Juri, P. N. (1979). *Tetrahedron Lett.* 407

Beckford, H. F., King, R. M. and Stoddart, J. F. (1978). *Tetrahedron Lett.* 171

Behr, J. P., Lehn, J. M. and Vierling, P. (1976). *J. C. S. Chem. Comm.* 621

Behr, J. P. and Lehn, J. M. (1978). *J. C. S. Chem. Comm.* 143

Belsky, I. (1977). *J. C. S. Chem. Comm.* 237

Blackmer, G. L., Nordyke, M. D., Vickrey, T. M., Bartsch, R. A. and Holwerda, R. A. (1978). *Inorg. Chem.* **17**, 3310

Bourgoin, M., Wong, K. H., Hui, J. Y. and Smid, J. (1975). *J. Amer. Chem. Soc.* **97**, 3462

Bradshaw, J. S. and Hui, J. Y. K. (1974). *J. Heterocyclic Chem.* **11**, 649

Bradshaw, J. S. (1978). *In* "Synthetic Multidentate Macrocyclic Compounds" (R. M. Izatt and J. J. Christensen, eds), p. 53. Academic Press, New York

Bradshaw, J. S., Maas, G. E., Izatt, R. M. and Christensen, J. J. (1979a). *Chem. Revs.* **79**, 37

Bradshaw, J. S., Maas, G. E., Izatt, R. M., Lamb, J. D. and Christensen, J. J. (1979b). *Tetrahedron Lett.* 635

Brändström, A. (1977). *Adv. Phys. Org. Chem.* **15**, 267.

Buncel, E. and Menon, B. (1977). *J. Amer. Chem. Soc.* **99**, 4457

Burden, I. J., Coxon, A. C., Stoddart, J. F. and Wheatley, C. M. (1977). *J. C. S. Perkin I* 220

Busch, D. H., Farmery, K., Goedken, V., Katovic, V., Melnyk, A. C., Sperati, C. R. and Tokel, N. (1971). *A. C. S. Adv. Chem. Ser.* **100**, 44

Busch, D. H. (1978). *Accounts Chem. Res.* **11**, 392

Bush, M. A. and Truter, M. R. (1970). *J. C. S. Chem. Comm.* 1439

Bush, M. A. and Truter, M. R. (1971). *J. Chem. Soc. B.* 1440

Cambillau, C., Sarthou, P. and Bram G. (1976). *Tetrahedron Lett.* 281

Cambillau, C., Bram, G., Corset, J., Riche, C. and Pascard-Billy, C. (1978). *Tetrahedron* **34**, 2675

Cardillo, G., Orena, M. and Sandri, S. (1976). *J. C. S. Chem. Comm.* 190

Chang, C. K. (1977). *J. Amer. Chem. Soc.* **99**, 2819

Chao, Y. and Cram, D. J. (1976). *J. Amer. Chem. Soc.* **98**, 1015

Chaput, G., Jeminet, G. and Juillard, J. (1975). *Can. J. Chem.* **53**, 2240

Chern, C., Di Cosimo, R., De Jesus, R. and San Fillippo, J. (1978). *J. Amer. Chem. Soc.* **100**, 7317

Chock, P. B. (1972). *Proc. Nat. Acad. Sci. USA* **69**, 1939

Chorev, M. and Klausner, Y. S. (1976). *J. C. S. Chem. Comm.* 596

Christensen, J. J., Hill, J. O. and Izatt, R. M. (1971). *Science* **174**, 459

Christensen, J. J., Eatough, D. J. and Izatt, R. M. (1974). *Chem. Revs.* **74**, 351

Cinquini, M., Montanari, F. and Tundo, P. (1975). *J. C. S. Chem. Comm.* 393

Cinquini, M. and Tundo, P. (1976). *Synthesis* 516

Cinquini, M., Colonna, S., Molinari, H., Montanari, F. and Tundo, P. (1976). *J. C. S. Chem. Comm.* 394

Cook, F. L., Bowers, C. W. and Liotta, C. L. (1974). *J. Org. Chem.* **39**, 3416

Corey, E. J., Nicolaou, K. C., Shibasaki, M., Machida, Y. and Shiner, C. S. (1975). *Tetrahedron Lett.* 3183

Cornelis, A., Laszlo, P. and Cambillau, C. (1978). *J. Chem. Res.* (*S*) 462

Cote, G. and Bauer, D. (1977). *J. Chem. Res.* (*M*) 3329

Cox, B. G., Hedwig, G. R., Parker, A. J. and Watts, D. W. (1974). *Aust. J. Chem.* **27**, 477

Cox, B. G., Schneider, H. and Stroka, J. (1978). *J. Amer. Chem. Soc.* **100**, 4746

Coxon, A. C. and Stoddart, J. F. (1975). *Carbohydrate Research* **44**, C 1

Coxon, A. C. and Stoddart, J. F. (1977). *J. C. S. Perkin I* 767

Coxon, A. C., Laidler, D. A., Pettman, R. B. and Stoddart, J. F. (1978). *J. Amer. Chem. Soc.* **100**, 8260

Cradwick, P. D. and Poonia, N. S. (1977). *Acta Cryst.* **B 33**, 197

Cram, D. J. and Cram, J. M. (1974). *Science* **183**, 803

Cram, D. J., Helgeson, R. C., Sousa, L. R., Timko, J. M., Newcomb, M., Moreau, P., De Jong, F., Gokel, G. W., Hoffman, D. H., Domeier, L. A., Peacock, S. C., Madan, K. and Kaplan, L. (1975). *Pure Appl. Chem.* **43**, 327

Cram, D. J. and Cram, J. M. (1978). *Accounts Chem. Res.* **11**, 8

Cram, D. J., Helgeson, R. C., Koga, K., Kyba, E. P., Madan, K., Sousa, L. R., Siegel, M. G., Moreau, P., Gokel, G. W., Timko, J. M. and Sogah, G. D. Y. (1978a). *J. Org. Chem.* **43**, 2758

Cram, D. J., Helgeson, R. C., Peacock, S. C., Kaplan, L. J., Domeier, L. A., Moreau, P., Koga, K., Mayer, J. M., Chao, Y., Siegel, M. G., Hoffman, D. H. and Sogah, G. D. Y. (1978b). *J. Org. Chem.* **43**, 1930

Curci, R. and Di Furia, F. (1975). *Int. J. Chem. Kinetics* **7**, 341

Curtis, W. D., Laidler, D. A., Stoddart, J. F. and Jones, G. H. (1975a). *J. C. S. Chem. Comm.* 833

Curtis, W. D., Laidler, D. A., Stoddart, J. F. and Jones, G. H. (1975b). *J. C. S. Chem. Comm.* 835

Curtis, W. D., King, R. M., Stoddart, J. F. and Jones, G. H. (1976). *J. C. S. Chem. Comm.* 284

Curtis, W. D., Laidler, D. A., Stoddart, J. F. and Jones, G. H. (1977a). *J. C. S. Perkin I* 1756

Curtis, W. D., Laidler, D. A., Stoddart, J. F., Wolstenholme, J. B. and Jones, G. H. (1977b). *Carbohydr. Res.* **57**, C 17

Dalley, N. K. (1978). *In* "Synthetic Multidentate Macrocyclic Compounds" (R. M. Izatt and J. J. Christensen, eds, p. 207. Academic Press, New York

Davidson, W. R. and Kebarle, P. (1976). *J. Amer. Chem. Soc.* **98**, 6133

Dechter, J. J. and Zink, J. I. (1976). *J. Amer. Chem. Soc.* **98**, 845

Dehmlow, E. V. and Lissel, M. (1976). *Tetrahedron Lett.* 1783

Dehmlow, E. V. (1977). *Angew. Chem.* **89**, 521

De Jong, F., Siegel, M. G. and Cram, D. J. (1975). *J. C. S. Chem. Comm.* 551

De Jong, F., Reinhoudt, D. N. and Smit, C. J. (1976a). *Tetrahedron Lett.* 1371

De Jong, F., Reinhoudt, D. N. and Smit, C. J. (1976b). *Tetrahedron Lett.* 1375

De Jong, F., Reinhoudt, D. N., Smit, C. J. and Huis, R. (1976c). *Tetrahedron Lett.* 4783

De Jong, F., Reinhoudt, D. N. and Huis, R. (1977). *Tetrahedron Lett.* 3985

De Jong, F., Reinhoudt, D. N. and Torny, G. J. (1979). *Tetrahedron Lett.* 911

Del Cima, F., Biggi, G. and Pietra, F. (1973). *J. C. S. Perkin II* 55

Di Biase, S. A. and Gokel, G. W. (1978). *J. Org. Chem.* **43**, 447

Dietrich, B. and Lehn, J. M. (1973). *Tetrahedron Lett.* 1225

Dietrich, B., Lehn, J. M. and Sauvage, J. P. (1973). *J. C. S. Chem. Comm.* 15

Dietrich, B., Lehn, J. M. and Simon, J. (1974). *Angew. Chem.* **86**, 443

Dorn, W. L., Knöchel, A., Oehler, J. and Rudolph, G. (1977). *Z. Naturforsch.* **B 32**, 776

Dou, H. J.-M., Gallo, R., Hassanaly, P. and Metzger, J. (1977). *J. Org. Chem.* **42**, 4275

Dupont Durst, H. (1974). *Tetrahedron Lett.* 2421

Dupont Durst, H., Zubrick, J. W. and Kieczykowski, G. R. (1974). *Tetrahedron Lett.* 1777

El-Kholy, A., Haanaes, E. and Rokstad, O. A. (1977). *Acta Chem. Scand.* **A 31**, 402

Evans, D. F., Wellington, S. L., Nadis, J. A. and Cussler, E. L. (1972). *J. Solution Chem.* **1**, 499

Evans, D. A. and Golob, A. M. (1975). *J. Amer. Chem. Soc.* **97**, 4765

Fedorynski, M. (1977), *Synthesis* 783

Fedorynski, M., Wojciechowski, K., Matacz, Z. and Makosza, M. (1978). *J. Org. Chem.* **43**, 4682

Fiandanese, V., Marchese, G., Naso, F. and Sciacovelli, O. (1973). *J. C. S. Perkin II* 1336

Fornasier, R., Montanari, F. and Podda, G. (1976). *Tetrahedron Lett.* 1381

Fraenkel, G. and Pechhold, E. (1970). *Tetrahedron Lett.* 153

Frensch, K. and Vögtle, F. (1977). *Tetrahedron Lett.* 2573

Frensdorf, H. K. (1971a). *J. Amer. Chem. Soc.* **93**, 600

Frensdorf, H. K. (1971b). *J. Amer. Chem. Soc.* **93**, 4684

Frimer, A. and Rosenthal, I. (1976). *Tetrahedron Lett.* 2809

Gandour, R. N., Walker, D. A., Nayak, A. and Newkome, G. R. (1978) *J. Amer. Chem. Soc.* **100**, 3608

Girodeau, J. M., Lehn, J. M. and Sauvage, J. P. (1975). *Angew. Chem.* **87**, 813
Gokel, G. W. and Cram, D. J. (1973). *J. C. S. Chem. Comm.* 481
Gokel, G. W., Cram, D. J., Liotta, C. L., Harris, H. P. and Cook, F. L. (1974). *J. Org. Chem.* **39**, 2445
Gokel, G. W., Timko, J. M. and Cram, D. J. (1975a). *J. C. S. Chem. Comm.* 444
Gokel, G. W., Timko, J. M. and Cram, D. J. (1975b). *J. C. S. Chem. Comm.* 394
Gokel, G. W. and Dupont Durst, H. (1976). *Synthesis* 168
Gokel, G. W., Gerdes, H. M. and Rebert, N. W. (1976). *Tetrahedron Lett.* 653
Gokel, G. W., Korzeniowski, S. H. and Blum, L. (1977). *Tetrahedron Lett.* 1633
Gold, V. and Sghibartz, C. M. (1978). *J. C. S. Chem. Comm.* 507
Goldberg, I. (1975). *Acta Cryst.* **B 31**, 754
Goldberg, I. (1977). *J. Amer. Chem. Soc.* **99**, 6494
Goldberg, I. (1978). *Acta Cryst.* **B 34**, 3387
Gordon, A. Z. and Rock, P. A. (1977). *J. Electrochem. Soc.* 534
Gramain, P. and Frère, Y. (1979). *Nouveau J. Chim.* **3**, 53
Grandjean, J., Laszlo, P., Vögtle, F. and Sieger, H. (1978). *Angew. Chem.* **90**, 902
Gutknecht, J., Schneider, H. and Stroka, J. (1978). *Inorg. Chem.* **17**, 3326
Gutmann, V. (1968). *In* "Coordination Chemistry in Nonaqueous Solvents" Springer Verlag, Vienna
Hain, W., Lehnert, R., Röttele, H. and Schröder, G. (1978). *Tetrahedron Lett.* 625
Handel, H. and Pierre, J. L. (1975a). *Tetrahedron* **31**, 2799
Handel, H. and Pierre, J. L. (1975b). *Tetrahedron* **31**, 997
Handel, H. and Pierre, J. L. (1976). *Tetrahedron Lett.* 741
Harris, C. J. (1977). *Sat. Heterocyclic Chem.* **4**, 229
Hartman, G. D. and Biffar, S. E. (1977). *J. Org. Chem.* **42**, 1468
Hautala, R. R. and Hastings, R. H. (1978). *J. Amer. Chem. Soc.* **100**, 648
Haymore, B. L., Ibers, J. A. and Meek, D. W. (1975). *Inorg. Chem.* **14**, 541
Haynes, D. H. and Pressman, B. C. (1974). *J. Membrane Biol.* **18**, 1
Hayward, R. C., Overton, C. H. and Whitham, G. H. (1976). *J. C. S. Perkin I* 2413
Helgeson, R. C., Timko, J. M. and Cram, D. J. (1973a). *J. Amer. Chem. Soc.* **95**, 3023
Helgeson, R. C., Koga, K., Timko, J. M. and Cram, D. J. (1973b). *J. Amer. Chem. Soc.* **95**, 3021
Helgeson, R. C., Timko, J. M., Moreau, P., Peacock, S. C., Mayer, J. M. and Cram, D. J. (1974). *J. Amer. Chem. Soc.* **96**, 6762
Helgeson, R. C., Tarnowski, T. L., Timko, J. M. and Cram, D. J. (1977). *J. Amer. Chem. Soc.* **99**, 6411
Herriott, A. W. and Picker, D. (1975). *J. Amer. Chem. Soc.* **97**, 2345
Hirao, A., Nakahama, S., Takahashi, M. and Yamazaki, N. (1978a). *Makromol. Chem.* **179**, 1735
Hirao, A., Nakahama, S., Takahashi, M. and Yamazaki, N. (1978b). *Makromol. Chem.* **179**, 915
Hirao, A., Nakahama, S., Takahashi, M., Mochizuki, H. and Yamazaki, Y. (1978c). *Makromol. Chem.* **179**, 2343
Hodgkinson, L. C., Leigh, S. J. and Sutherland, I. O. (1976a). *J. C. S. Chem. Comm.* 639
Hodgkinson, L. C., Leigh, S. J. and Sutherland, I. O. (1976b). *J. C. S. Chem. Comm.* 640
Hofmanova, A., Koryta, J., Brezina, M. and Mittal, M. L. (1978). *Inorg. Chim. Acta* **28**, 73
Hogen-Esch, T. E. (1977). *Adv. Phys. Org. Chem.* **15**, 154.
Horner, L. and Brich, W. (1978). *Chem. Ber.* **111**, 574
Hughes, D. L. and Wingfield, J. N. (1977). *J. C. S. Chem. Comm.* 804
Hunter, D. H. and Shearing, D. J. (1971). *J. Amer. Chem. Soc.* **93**, 2348
Hunter, D. H., Lee, W. and Kim, S. K. (1974). *J. C. S. Chem. Comm.* 1018

Hunter, D. H. and Perry, R. A. (1977). *Synthesis* 37

Hunter, D. H., Hamity, M., Patel, V. and Perry, R. A. (1978). *Can J. Chem.* **56**, 104

Idemori, K., Takagi, M. and Matsuda, T. (1977). *Bull. Chem. Soc. Jap.* **50**, 1355

Ishizu, K., Kohama, H. and Mukai, K. (1978). *Chem. Lett.* 227

Iwachido, T., Kimura, M. and Toei, K. (1976). *Chem. Lett.* 1101

Iwachido, T., Minami, M., Sadakane, A. and Toei, K. (1977). *Chem. Lett.* 1511

Iwachido, T., Sadakane, A. and Toei, K. (1978). *Bull. Chem. Soc. Jap.* **51**, 629

Izatt, R. M., Nelson, D. P., Rytting, J. H., Haymore, B. L. and Christensen, J. J. (1971). *J. Amer. Chem. Soc.* **93**, 1619

Izatt, R. M., Eatough, D. J. and Christensen, J. J. (1973). *Structure and Bonding* **16**, 161

Izatt, R. M., Terry, R. E., Haymore, B. L., Hansen, L. D., Dalley, N. K., Avondet, A. G. and Christensen, J. J. (1976a). *J. Amer. Chem. Soc.* **98**, 7620

Izatt, R. M., Terry, R. E., Nelson, D. P., Chan, Y., Eatough, D. J., Bradshaw, J. S., Hansen, L. D. and Christensen, J. J. (1976b). *J. Amer. Chem. Soc.* **98**, 7626

Izatt, R. M., Lamb, J. D., Maas, G. E., Asay, R. E., Bradshaw, J. S. and Christensen, J. J. (1977a). *J. Amer. Chem. Soc.* **99**, 2365

Izatt, R. M., Lamb, J. D., Asay, R. E., Maas, G. E., Bradshaw, J. S., Christensen, J. J. and Moore, S. S. (1977b). *J. Amer. Chem. Soc.* **99**, 6134

Izatt, R. M., Lamb, J. D., Christensen, J. J. and Haymore, B. L. (1977c). *J. Amer. Chem. Soc.* **99**, 8344

Izatt, R. M., Terry, R. E., Hansen, L. D., Avondet, A. G., Bradshaw, J. S., Dalley, N. K., Jensen, T. E. and Christensen, J. J. (1978a) *Inorg. Chim. Acta* **30**, 1

Izatt, R. M., Izatt, N. E., Rossiter, B. E., Christensen, J. J. and Haymore, B. L. (1978b). *Science* **199**, 994

Izatt, R. M., Lamb, J. D., Rossiter, B. E., Izatt, N. E., Christensen, J. J. and Haymore, B. L. (1978c). *J. C. S. Chem. Comm.* 386

Jaber, A. M. Y., Moody, G. J. and Thomas, J. D. R. (1977). *J. Inorg. Nucl. Chem.* **39**, 1689

Jawaid, M. and Ingman, F. (1978). *Talanta* **25**, 91

Jepson, B. E. and De Witt, R. (1976). *J. Inorg. Nucl. Chem.* **38**, 1175

Johnson, R. A. and Nidy, E. G. (1975). *J. Org. Chem.* **40**, 1680

Johnson, R. A., Nidy, E. G. and Merritt, M. V. (1978). *J. Amer. Chem. Soc.* **100**, 7960

Juri, P. N. and Bartsch, R. A. (1979). *J. Org. Chem.* **44**, 143

Kappenstein, C. (1974). *Bull. Soc. Chim. Fr.* 89

Kaufmann, E., Lehn, J. M. and Sauvage, J. P. (1976). *Helv. Chim. Acta* **59**, 1099

Kaufmann, R., Knöchel, A., Kopf, J., Oehler, J. and Rudolph, G. (1977). *Chem. Ber.* **110**, 2249

Kawashima, N., Kawashima, T., Otsubo, T. and Misumi, S. (1978). *Tetrahedron Lett.* 5025

Kime, D. E. and Norymberski, J. K. (1977). *J. C. S. Perkin I* 1048

Kimura, K., Maeda, T. and Shono, T. (1978). *Anal. Letters* **A 11**, 821

Klausner, Y. S. and Chorev, M. (1977). *J. C. S. Perkin I*, 627

Knipe, A. C., Sridhar, N. and Loughran, A. (1976). *J. C. S. Chem. Comm.* 630

Knöchel, A., Rudolph, G. and Thiem, J. (1974). *Tetrahedron Lett.* 551

Knöchel, A. and Rudolph, G. (1974). *Tetrahedron Lett.* 3739

Knöchel, A., Oehler, J. and Rudolph, G. (1975). *Tetrahedron Lett.* 3167

Knöchel, A., Oehler, J. and Rudolph, G. (1977a). *Z. Naturforsch.* **B 32**, 783

Knöchel, A., Oehler, J., Rudolph, G. and Sinnwell, V. (1977b). *Tetrahedron* **33**, 119

Knöchel, A., Kopf, J., Oehler, J. and Rudolph, G. (1978). *J. C. S. Chem. Comm.* 595

Kodama, M. and Kimura, E. (1976). *Bull. Chem. Soc. Jap.* **49**, 2465

Koenig, K. E., Helgeson, R. C. and Cram, D. J. (1976). *J. Amer. Chem. Soc.* **98**, 4018

Kopolow, S., Machacek, Z., Takaki, U. and Smid, J. (1973). *J. Macromol. Sci.-Chem.* **A 7**. 1015

Korzeniowski, S. H., Petcavich, R. J., Coleman, M. M. and Gokel, G. W. (1977). *Tetrahedron Lett.* 2647

Kostikov, R. R. and Molchanov, A. P. (1975). *Zh. Org. Khimii*, **11**, 1767

Krespan, C. G. (1974). *J. Org. Chem.* **39**, 2351

Kurts, A. L., Demyanov, P. I., Beletskaya, I. P. and Rentov, O. A. (1973). *Zh. Org. Khimii.* **9**, 1313

Kurts, A. L., Sakembawa, S. M., Beletskaya, I. P. and Rentov, O. A. (1974). *Zh. Org. Khimii.* **10**, 1572

Kwon, S., Nishimura, Y., Ikeda, M. and Tamura, Y. (1976). *Synthesis* 249

Kyba, E. P., Siegel, M. G., Sousa, L. R., Sogah, G. D. Y. and Cram, D. J. (1973a). *J. Amer. Chem. Soc.* **95**, 2691

Kyba, E. P., Koga, K., Sousa, L. R., Siegel, M. G. and Cram, D. J. (1973b). *J. Amer. Chem. Soc.* **95**, 2692

Kyba, E. P., Gokel, G. W., De Jong, F., Koga, K., Sousa, L. R., Siegel, M. G., Kaplan, L., Sogah, G. D. Y. and Cram, D. J. (1977a). *J. Org. Chem.* **42**, 4173

Kyba, E. P., Helgeson, R. C., Madan, K., Gokel, G. W., Tarnowski, T. L., Moore, S. S. and Cram, D. J. (1977b). *J. Amer. Chem. Soc.* **99**, 2564

Kyba, E. P., Timko, J. M., Kaplan, L. J., De Jong, F., Gokel, G. W. and Cram, D. J. (1978). *J. Amer. Chem. Soc.* **100**, 4555.

Laidler, D. A. and Stoddart, J. F. (1976). *J. C. S. Chem. Comm.* 979

Laidler, D. A. and Stoddart, J. F. (1977a). *J. C. S. Chem. Comm.* 481

Laidler, D. A. and Stoddart, J. F. (1977b). *Carbohydr. Res.* **55**, C 1

Laidler, D. A. and Stoddart, J. F. (1979). *Tetrahedron Lett.* 453

Laidler, D. A., Stoddart, J. F. and Wolstenholme, J. B. (1979). *Tetrahedron Lett.* 465

Landini, D., Montanari, F. and Pirisi, F. M. (1974). *J. C. S. Chem. Comm.* 879

Landini, D., Maia, A. M., Montanari, F. and Pirisi, F. M. (1975). *Gazz. Chim. Ital.* **105**, 863

Landini, D., Maia, A. and Montanari, F. (1978). *J. Amer. Chem. Soc.* **100**, 2796

Laureillard, J. and Casadevall, E. (1977). *Tetrahedron Lett.* 2875

Lee, D. G. and Chang, V. S. (1978). *J. Org. Chem.* **43**, 1532

Lefour, J. M. and Loupy, A. (1978). *Tetrahedron*, **34**, 2597

Lehmkuhl, H., Rabet, F. and Hauschild, K. (1977). *Synthesis* 184

Lehn, J. M. and Sauvage, J. P. (1971). *Chem. Comm.* 440

Lehn, J. M. (1973). *Structure and Bonding* **16**, 1

Lehn, J. M. and Sauvage, J. P. (1975). *J. Amer. Chem. Soc.* **97**, 6700

Lehn, J. M. and Simon, J. (1977). *Helv. Chim. Acta* **60**, 141

Lehn, J. M. (1978). *Accounts Chem. Res.* **11**, 49

Lehn, J. M., Simon, J. and Moradpour, A. (1978). *Helv. Chim. Acta* **61**, 2407

Lehn, J. M. and Sirlin, C. (1978). *J. C. S. Chem. Comm.* 949

Leigh, S. J. and Sutherland, I. O. (1975). *J. C. S. Chem. Comm.* 414

Lemmetyinen, H., Edelmann, K. and Koskikallio, J. (1978). *Finn. Chem. Lett.* 169

Liesegang, G. W., Farrow, M. M., Purdie, N. and Eyring, E. M. (1976). *J. Amer. Chem. Soc.* **98**, 6905

Liesegang, G. W., Farrow, M. M., Vasquez, F. A., Purdie, N. and Eyring, E. M. (1977). *J. Amer. Chem. Soc.* **99**, 3240

Liesegang, G. W., Farrow, M. M., Rodriguez, L. J., Burnham, R. K. and Eyring, E. M. (1978). *Int. J. Chem. Kin.* **10**, 471

Liesegang, G. W. and Eyring, E. M. (1978). *In* "Synthetic Multidentate Macrocyclic Compounds" (R. M. Izatt and J. J. Christensen, eds), p. 245. Academic Press, New York

Liotta, C. L. and Harris, H. P. (1974). *J. Amer. Chem. Soc.* **96**, 2250

Liotta, C. L., Harris, H. P., McDermott, M., Gonzalez, T. and Smith, K. (1974). *Tetrahedron Lett.* 2417
Liotta, C. L. and Grisdale, E. E. (1975). *Tetrahedron Lett.* 4205
Liotta, C. L., Dabdoub, A. M. and Zalkow, L. H. (1977). *Tetrahedron Lett.* 1117
Liotta, C. L. (1978). *In* "Synthetic Multidentate Macrocyclic Compounds" (R. M. Izatt and J. J. Christensen, eds), p. 111. Academic Press, New York
Litvak, V. V. and Shein, S. M. (1976). *J. Org. Chem. USSR* **12**, 1693
Live, D. and Chan, S. I. (1976). *J. Amer. Chem. Soc.* **98**, 3796
Loupy, A., Seyden-Penne, J. and Tchoubar, B. (1976). *Tetrahedron Lett.* 1677
Loupy, A. and Seyden-Penne, J. (1978). Tetrahedron Lett. 2571
Mack, M. P., Hendrixson, R. R., Palmer, R. A. and Ghirardelli, R. G. (1976). *J. Amer. Chem. Soc.* **98**, 7830
Makosza, M. and Ludwikow, M. (1974). *Angew. Chem. Int.* **13**, 665
Makosza, M. (1975). *Pure Appl. Chem.* **43**, 439
Mallinson, P. R. and Truter, M. R. (1972). *J. C. S. Perkin II* 1818
Mandolini, L. and Masci, B. (1977). *J. Amer. Chem. Soc.* **99**, 7709
Marcus, Y. and Asher, L. E. (1978), *J. Phys. Chem.* **82**, 1246
Mariani, C., Modena, G. and Scorrano, G. (1978). *J. Chem. Res.* (*M*) 4601
Markezich, R. L., Zamek, O. S., Donahne, P. E. and Williams, F. J. (1977). *J. Org. Chem.* **42**, 3435
Martin, J. C. and Bloch, D. R. (1971). *J. Amer. Chem. Soc.* **93**, 451
Maskornick, M. J. (1972). *Tetrahedron Lett.* 1797
Matsuda, T. and Koida, K. (1973). *Bull. Chem. Soc. Jap.* **46**, 2259
Matsui, T. and Koga, K. (1978). *Tetrahedron Lett.* 1115
Matsuura, N., Umemoto, K., Takeda, Y. and Sasaki, A. (1976). *Bull. Chem. Soc. Jap.* **49**, 1246
McKenzie, W. M. and Sherrington, D. C. (1978). *J. C. S. Chem. Comm.* 541
McKervey, M. A. and Mulholland, D. L. (1977). *J. C. S. Chem. Comm.* 438
McLachlan, R. D. (1974). *Spectrochimica Acta* **30 A**, 2153
Mei, E., Dye, J. L. and Popov, A. I. (1977a). *J. Amer. Chem. Soc.* **99**, 5308
Mei, E., Popov, A. I. and Dye, J. L. (1977b). *J. Phys. Chem.* **81**, 1677
Metcalfe, J. C., Stoddart, J. F. and Jones, G. (1977). *J. Amer. Chem. Soc.* **99**, 8317
Mikolajczyk, M., Grzejszczak, S., Midura, W. and Zatorski, A. (1975). *Synthesis* 278
Molinari, H., Montanari, F. and Tundo, P. (1977). *J. C. S. Chem. Comm.* 639
Moore, S. S., Tarnovski, T. L., Newcomb, M. and Cram, D. J. (1977). *J. Amer. Chem. Soc.* **99**, 6398
Moro-Oka, Y., Chung, P. J., Arakawa, H. and Ikawa, T. (1976). *Chem. Lett.* 1293
Moss, R. A. and Pilkiewicz, F. G. (1974). *J. Amer. Chem. Soc.* **96**, 5623
Moss, R. A., Joyce, M. A. and Pilkiewicz, F. G. (1975). *Tetrahedron Lett.* 2425
Movsumzade, M. M., Shabanov, A. L., Agaev, F. K. and Abdullabekov, I. M. (1976). *Zh. Org. Khimii* **12**, 2400
Nakabayashi, T., Kawamura, S., Horii, T. and Hamada, M. (1976). *Chem. Lett.* 869
Nakajima, Y., Kinishi, R., Oda, J. and Inouye, Y. (1977). *Bull. Chem. Soc. Jap.* **50**, 2025
Naso, F. and Ronzini, L. (1974). *J. C. S. Perkin I* 340
Née, G. and Tchoubar, B. (1976). *C. R. Acad. Sc. Paris* **283**, C-223
Newcomb, G. M., Helgeson, R. C. and Cram, D. J. (1974). *J. Amer. Chem. Soc.* **96**, 7367
Newcomb, M., Moore, S. S. and Cram, D. J. (1977a). *J. Amer. Chem. Soc.* **99**, 6405
Newcomb, M., Timko, J. M., Walba, D. M. and Cram, D. J. (1977b). *J. Amer. Chem. Soc.* **99**, 6392
Newkome, G. R., Sauer, J. D., Roper, J. M. and Hagar, D. C. (1977). *Chem. Rev.* **77**, 513
Noe, E. A. and Raban, M. (1974). *J. Amer. Chem. Soc.* **96**, 6184

Nordlander, J. E., Catalane, D. B., Eberlein, T. H., Farkas, L. V., Howe, R. S., Stevens, R. M. and Tripoulas, N. A. (1978). *Tetrahedron Lett.* 4987
Normant, H., Cuvigny, T. and Savignac, P. (1975). *Synthesis* 805
Ohtomi, M., Yoneyama, S., Arai, K. and Akabori, S. (1976). *Nippon Kagaku Kaishi* 1878
Padwa, A. and Dehm, D. (1975). *J. Org. Chem.* **40**, 3139
Pankova, M. and Zavada, J. (1977). *Coll. Czech. Chem. Comm.* **42**, 1981
Pannell, K. H., Yee, W., Lewandos, G. S. and Hambrick, D. C. (1977). *J. Amer. Chem. Soc.* **99**, 1457
Pannell, K. H. and McIntosh, J. (1977). *Chem. Ind.*, 872
Parsons, D. G. (1978). *J. C. S. Perkin I* 451
Peacock, S. C. and Cram, D. J. (1976). *J. C. S. Chem. Comm.* 282
Peacock, S. C., Domeier, L. A., Gaeta, F. C. A., Helgeson, R. C., Timko, J. M. and Cram, D. J. (1978). *J. Amer. Chem. Soc.* **100**, 8190
Peacock, S. C., Walba, D. M., Gaeta, F. C. A., Helgeson, R. C. and Cram, D. J. (1980). *J. Amer. Chem. Soc.* **102**, 2043
Pedersen, C. J. (1967a). *J. Amer. Chem. Soc.* **89**, 2495
Pedersen, C. J. (1967b). *J. Amer. Chem. Soc.* **89**, 7017
Pedersen, C. J. (1971) *J. Org. Chem.* **36**, 1690
Pedersen, C. J. and Frensdorf, H. K. (1972). *Angew. Chem. Int.* **11**, 16
Pedersen, C. J. (1978). *In* "Synthetic Multidentate Macrocyclic Compounds" (R. M. Izatt and J. J. Christensen, eds), p. 1. Academic Press, New York
Pettman, R. B. and Stoddart, J. F. (1979a). *Tetrahedron Lett.* 457
Pettman, R. B. and Stoddart, J. F. (1979b). *Tetrahedron Lett.* 461
Pierre, J. L. and Handel, H. (1974). *Tetrahedron Lett.* 2317
Pierre, J. L., Handel, H. and Perrand, R. (1975). *Tetrahedron* **31**, 2795
Pierre, J. L., Le Goaller, R., Handel, H. and Ripert, A. (1978a). *Tetrahedron Lett.* 3259
Pierre, J. L., Le Goaller, R. and Handel, H. (1978b). *J. Amer. Chem. Soc.* **100**, 8021
Platzner, T., Thomas, J. K. and Grätzel, M. (1978). *Z. Naturforsch.* **33 B**, 614
Popov, A. I. (1976). *In* "Solute-Solvent Interactions" (J. F. Coetzee and C. D. Ritchie, eds), Vol. 2, p. 271. Marcel Dekker Inc., New York
Prestegard, J. H. and Chan, S. I. (1970). *J. Amer. Chem. Soc.* **92**, 4440
Raban, M., Noe, E. A. and Yamamoto, G. (1977). *J. Amer. Chem. Soc.* **99**, 6527
Rall, G. J. H., Oberholzer, M. E., Ferreira, D. and Roux, D. G. (1976). *Tetrahedron Lett.* 1033
Rasmussen, J. K. (1977). *Chem. Lett.* 1295
Rasshofer, W. and Vögtle, F. (1978). *Tetrahedron Lett.* 309
Regen, S. L. (1975). *J. Amer. Chem. Soc.* **97**, 5956
Regen, S. L. (1976). *J. Amer. Chem. Soc.* **98**, 6270
Regen, S. L. (1977). *J. Org. Chem.* **42**, 875
Regen, S. L. and Nigam, A. (1978). *J. Amer. Chem. Soc.* **100**, 7773
Regen, S. L., Nigam, A. and Besse, J. J. (1978). *Tetrahedron Lett.* 2757
Reinhoudt, D. N., Gray, R. T., Smit, C. J. and Veenstra, I. (1976). *Tetrahedron* **32**, 1161
Reinhoudt, D. N., Gray, R. T., De Jong, F. and Smit, C. J. (1977). *Tetrahedron* **33**, 563
Reinhoudt, D. N. and De Jong, F. (1979). *In* "Progress in Macrocyclic Chemistry" (R. M. Izatt and J. J. Christensen, eds), Vol. 1, p. 157. Wiley, New York
Rodriguez, L. J., Liesegang, G. W., White, R. D., Farrow, M. M., Purdie, N. and Eyring, E. M. (1977). *J. Phys. Chem.* **81**, 2118
Rodriguez, L. J., Liesegang, G. W., Farrow, M. M., Purdie, N. and Eyring, E. M. (1978). *J. Phys. Chem.* **82**, 647
Roitman, J. N., and Cram, D. J. (1971). *J. Amer. Chem. Soc.* **93**, 2231
Russell, G. A., Wallraf, G. and Gerlock, J. L. (1978). *J. Phys. Chem.* **82**, 1161

Ryba, O. and Petranek, J. (1973). *Electroanal. Chem. Interf. Electrochem.* **44**, 423
Sadakane, A., Iwachido, T. and Toei, K. (1975). *Bull. Chem. Soc. Jap.* **48**, 60
Sam, D. J. and Simmons, H. E. (1972). *J. Amer. Chem. Soc.* **94**, 4024
Sam, D. J. and Simmons, H. E. (1974). *J. Amer. Chem. Soc.* **96**, 2252
San Fillippo, J., Chern, C. and Valentine, J. S. (1975). *J. Org. Chem.* **40**, 1678
San Fillippo, J., Romano, L. J., Chern, C. and Valentine, J. S. (1976). *J. Org. Chem.* **41**, 586
Sasaki, T., Eguchi, S., Ohno, M. and Nakata, F. (1976). *J. Org.Chem.* **41**, 2408
Sasaki, T., Eguchi, S. and Nakata, F. (1978), *Tetrahedron Lett.* 1999
Schlosser, M., Spahic, B., Tarchini, C. and Van Chau, L. (1975). *Angew. Chem.* **87**, 346
Scully, F. E. and Davis, R. C. (1978). *J. Org. Chem.* **43**, 1467
Sekine, T., Wakabayashi, H. and Hasegawa, Y. (1978). *Bull. Chem. Soc. Jap.* **51**, 645
Sepp, D. T., Scherer, K. V. and Weber, W. P. (1974). *Tetrahedron Lett.* 2983
Shaw, J. E., Hsia, D. Y., Parries, G. S. and Sawyer, T. K. (1978). *J. Org. Chem.* **43**, 1017
Shchori, E., Jagur-Grodzinski, J., Luz, Z. and Shporer, M. (1971). *J. Amer. Chem. Soc.* **93**, 7133
Shchori, E. and Jagur-Grodzinski, J. (1973). *Israel J. Chem.* **11**, 243
Shchori, E., Jagur-Grodzinski, J. and Shporer, M. (1973). *J. Amer. Chem. Soc.* **95**, 3842
Shchori, E., Nae, N. and Jagur-Grodzinski, J. (1975). *J. C. S. Dalton* 2381
Shporer, M. and Luz, Z. (1975). *J. Amer. Chem. Soc.* **97**, 665
Smid, J. (1972). *Angew. Chem. Int.* **11**, 112
Smid, J. (1976). *Pure Appl. Chem.* **48**, 343
Smith, S. G. and Hanson, M. P. (1971). *J. Org. Chem.* **36**, 1931
Sogah, G. D. Y. and Cram, D. J. (1975). *J. Amer. Chem. Soc.* **97**, 1259
Sogah, G. D. Y. and Cram, D. J. (1976). *J. Amer. Chem. Soc.* **98**, 3038
Sogah, G. D. Y. and Cram, D. J. (1979). *J. Amer. Chem. Soc.* **101**, 3035
Sousa, L. R., Hoffman, D. H., Kaplan, L. and Cram, D. J. (1974). *J. Amer. Chem. Soc.* **96**, 7100
Sousa, L. R., Sogah, G. D. Y., Hoffman, D. H. and Cram, D. J. (1978). *J. Amer. Chem. Soc.* **100**, 4569
Srivanavit, C., Zink. J. I. and Dechter, J. J. (1977). *J. Amer. Chem. Soc.* **99**, 5876
Starks, C. M. and Owens, R. M. (1973). *J. Amer. Chem. Soc.* **95**, 3613
Starks, C. M. and Liotta, C. (1978). "Phase Transfer Catalysis". Academic Press, New York
Stoddart, J. F. and Wheatley, C. M. (1974). *J. C. S. Chem. Comm.* 390
Suh, I. H. and Saenger, W. (1978). *Angew. Chem.* **90**, 565
Svoboda, M., Hapala, J. and Zavada, J. (1972). *Tetrahedron Lett.* 265
Takagi, M., Tazaki, M. and Ueno, K. (1978). *Chem. Lett.* 1179
Takayama, K., Nambu, N. and Nagai, T. (1977). *Chem. Pharm. Bull. Jap.* **25**, 2608
Tarnowski, T. L. and Cram, D. J. (1976). *J. C. S. Chem. Comm.* 661
Thomassen, L. M., Ellingsen, T. and Ugelstad, J. (1971). *Acta Chem. Scand.* **25**, 3024
Timko, J. M. and Cram, D. J. (1974). *J. Amer. Chem. Soc.* **96**, 7159
Timko, J. M., Helgeson, R. C., Newcomb, M., Gokel, G. W. and Cram, D. J. (1974). *J. Amer. Chem. Soc.* **96**, 7097
Timko, J. M., Moore, S. S., Walba, D. M., Hiberty, P. C. and Cram, D. J. (1977). *J. Amer. Chem. Soc.* **99**, 4207
Timko, J. M., Helgeson, R. C. and Cram, D. J. (1978). *J. Amer. Chem. Soc.* **100**, 2828
Tosteson, D. C. (1968). *Fed. Proc.* **27**, 1269
Truter, M. R. (1973). *Structure and Bonding* **16**, 71
Tümmler, B., Maass, G., Weber, E., Wehner, W. and Vögtle, F. (1977). *J. Amer. Chem. Soc.* **99**, 4683
Ungaro, R., El Haj, B. and Smid, J. (1976). *J. Amer. Chem. Soc.* **98**, 5198
Valentine, J. S. and Curtis, A. B. (1975). *J. Amer. Chem. Soc.* **97**, 224
Van Bergen, T. J. and Kellogg, R. M. (1976). *J. Amer. Chem. Soc.* **98**, 1962

Van Bergen, T. J. and Kellogg, R. M. (1977). *J. Amer. Chem. Soc.* **99**, 3882

Van der Veen, R. H., Kellogg, R. M., Vos, A. and Van Bergen, T. J. (1978). *J. C. S. Chem. Comm.* 923

Vander Zwan, M. C. and Hartner, F. W. (1978). *J. Org. Chem.* **43**, 2655

Vögtle, F. and Neumann, P. (1973). *Chem. Z.* **97**, 600

Vögtle, F. and Weber, E. (1974). *Angew. Chem. Int.* **13**, 814

Wakabayashi, T. and Kato, Y. (1977). *Tetrahedron Lett.* 1235

Weber, W. P. and Gokel, G. W. (1977). "Phase Transfer Catalysis in Organic Synthesis" Springer Verlag, Berlin

Wester, N. and Vögtle, F. (1978). *J. Chem. Res.* (*M*) 4856

Wiegers, K. E. and Smith, S. G. (1978). *J. Org. Chem.* **43**, 1126

Wong, K. H., Konizer, G. and Smid, J. (1970). *J. Amer. Chem. Soc.* **92**, 666

Wong, K.-H., Bourgoin, M. and Smid, J. (1974). *J. C. S. Chem. Comm.* 715

Wong, L. and Smid, J. (1977). *J. Amer. Chem. Soc.* **99**, 5637

Wong, K.-H. (1978). *J. C. S. Chem. Comm.* 282

Wudl, F. and Gaeta, F. (1972). *J. C. S. Chem. Comm.* 107

Yamaguchi, T. and Van der Plas, H. C. (1977). *Rec. Trav. Chim. Pays-Bas* **96**, 89

Yamamura, K. and Murahashi, S. (1977). *Tetrahedron Lett.* 4429

Yanagida, S., Noji, Y. and Okahara, M. (1977a). *Tetrahedron Lett.* 2893

Yanagida, S., Noji, Y. and Okahara, M. (1977b). *Tetrahedron Lett.* 2337

Yanagida, S., Takahashi, K. and Okahara, M. (1977c). *Bull. Chem. Soc. Jap.* **50**, 1386

Ykman, P. and Hall, H. K. (1975). *Tetrahedron Lett.* 2429

Zaug, H. E., Ratajczyk, J. F., Leonard, J. E. and Schaefer, A. D. (1972). *J. Org. Chem.* **37**, 2249

Zavada, J., Svoboda, M. and Pankova, M. (1972). *Tetrahedron Lett.* 711

Zavada, J., Pankova, M., Svoboda, M. and Schlosser, M. (1973). *J. C. S. Chem. Comm.* 168

Zavada, J., Pankova, M. and Svoboda, M. (1976). *Coll. Czech. Chem. Comm.* **41**, 3778

Zavada, J. and Pankova, M. (1977). *Coll. Czech. Chem. Comm.* **42**, 3421

Zavada, J. and Pankova, M. (1978). *Coll. Czech. Chem. Comm.* **43**, 1080

Zubrick, J. W., Dunbar, B. I. and Dupont Durst, H. (1975). *Tetrahedron Lett.* 71

Subject Index

D

E